D1072059

DESIGN ASSURANCE
FOR ENGINEERS AND MANAGERS

MECHANICAL ENGINEERING

A Series of Textbooks and Reference Books

EDITORS

L. L. FAULKNER

Department of Mechanical Engineering
The Ohio State University
Columbus, Ohio

S. B. MENKES

Department of Mechanical Engineering
The City College of the
City University of New York
New York, New York

1. Spring Designer's Handbook, *by Harold Carlson*
2. Computer-Aided Graphics and Design, *by Daniel L. Ryan*
3. Lubrication Fundamentals, *by J. George Wills*
4. Solar Engineering for Domestic Buildings, *by William A. Himmelman*
5. Applied Engineering Mechanics: Statics and Dynamics, *by G. Boothroyd and C. Poli*
6. Centrifugal Pump Clinic, *by Igor J. Karassik*
7. Computer-Aided Kinetics for Machine Design, *by Daniel L. Ryan*
8. Plastics Products Design Handbook, Part A: Materials and Components; Part B: Processes and Design for Processes, *edited by Edward Miller*
9. Turbomachinery: Basic Theory and Applications, *by Earl Logan, Jr.*
10. Vibrations of Shells and Plates, *by Werner Soedel*
11. Flat and Corrugated Diaphragm Design Handbook, *by Mario Di Giovanni*
12. Practical Stress Analysis in Engineering Design, *by Alexander Blake*
13. An Introduction to the Design and Behavior of Bolted Joints, *by John H. Bickford*
14. Optimal Engineering Design: Principles and Applications, *by James N. Siddall*
15. Spring Manufacturing Handbook, *by Harold Carlson*

16. Industrial Noise Control: Fundamentals and Applications, *edited by Lewis H. Bell*

17. Gears and Their Vibration: A Basic Approach to Understanding Gear Noise, *by J. Derek Smith*

18. Chains for Power Transmission and Material Handling: Design and Applications Handbook, *by the American Chain Association*

19. Corrosion and Corrosion Protection Handbook, *edited by Philip A. Schweitzer*

20. Gear Drive Systems: Design and Application, *by Peter Lynwander*

21. Controlling In-Plant Airborne Contaminants: Systems Design and Calculations, *by John D. Constance*

22. CAD/CAM Systems Planning and Implementation, *by Charles S. Knox*

23. Probabilistic Engineering Design: Principles and Applications, *by James N. Siddall*

24. Traction Drives: Selection and Application, *by Frederick W. Heilich III and Eugene E. Shube*

25. Finite Element Methods: An Introduction, *by Ronald L. Huston and Chris E. Passerello*

26. Mechanical Fastening of Plastics: An Engineering Handbook, *by Brayton Lincoln, Kenneth J. Gomes, and James F. Braden*

27. Lubrication in Practice, Second Edition, *edited by W. S. Robertson*

28. Principles of Automated Drafting, *by Daniel L. Ryan*

29. Practical Seal Design, *edited by Leonard J. Martini*

30. Engineering Documentation for CAD/CAM Applications, *by Charles S. Knox*

31. Design Dimensioning with Computer Graphics Applications, *by Jerome C. Lange*

32. Mechanism Analysis: Simplified Graphical and Analytical Techniques, *by Lyndon O. Barton*

33. CAD/CAM Systems: Justification, Implementation, Productivity Measurement, *by Edward J. Preston, George W. Crawford, and Mark E. Coticchia*

34. Steam Plant Calculations Manual, *by V. Ganapathy*

35. Design Assurance for Engineers and Managers, *by John A. Burgess*

OTHER VOLUMES IN PREPARATION

DESIGN ASSURANCE
FOR ENGINEERS AND MANAGERS

John A. Burgess
Westinghouse Electric Corporation
Power Transformer Plant
Muncie, Indiana

MARCEL DEKKER, INC.   New York and Basel

Library of Congress Cataloging in Publication Data

Burgess, John A., [date]
    Design assurance for engineers and managers.

    (Mechanical engineering ; 35)
    Bibliography: p.
    Includes index.
    1. Engineering design. 2. Reliability (Engineering)
I. Title. II. Series.
TA174.B87 1984   620'.00425   84-20047
ISBN 0-8247-7258-X

MARCEL DEKKER, INC.
270 Madison Avenue, New York, New York 10016

Current printing (last digit):
10 9 8 7 6 5 4

PRINTED IN THE UNITED STATES OF AMERICA

To my wife, Alma

Preface

People in industry and government have long recognized the major impact that design has on products and their usefulness. Yet, in many cases, only limited attention is given to the quality of the design effort. It is often just assumed that the engineering work is good. Even though much has been written about quality control, nearly all of it has been directed towards manufacturing and production. Little has been written about the application of quality methods to engineering activities.

This book is written for engineers and managers everywhere who want to achieve high levels of quality in their design work. It describes the concepts and methods of a discipline called *design assurance*. The book will also be useful to quality assurance professionals who have always wanted to know more about the engineering process but were afraid to ask.

There is another audience for this book, too. It can be used as supplemental reading for graduate and undergraduate engineering students. It reveals many nontechnical aspects that are necessary for getting the work done in an engineering department. Typically, these matters come as a surprise to the new graduate. Accordingly, this book can aid in the transition from student to practicing engineer.

Although this book describes several concepts that may be new to the reader, it presents them in ways engineers can use them in their daily activities. The various chapters contain recommendations, examples, and preferred practices that illustrate how the concepts can be applied to many products and many industries. The methods presented in this book typically provide a series of checks and balances in the engineering process. Most of these are accomplished within the engineering department itself. In fact, some readers will simply consider these methods as "good engineering practices."

The contents of this book are based largely on the author's observations and experiences with many companies over the past 30 years. Having worked as an engineer and manager in several engineering and quality assurance departments, I have seen these practices from both vantage points. Some of the techniques are well known and widely recognized. Others are more subtle and less visible to the casual observer. However, collectively, they represent a substantial part of that body of knowledge known as design assurance.

The value of such methods comes in their application. Yet, as they say, there is more than one way to skin the cat. I encourage the readers to experiment with these techniques. Modify them to fit your local situation, but, above all, give them a try.

As you can imagine, a book like this doesn't happen overnight. The concepts grow from ideas and events over a long period of time.

I owe much to my compatriots at Westinghouse for their knowledge, imagination, and persistence in developing, honing, and applying many of the tools and techniques discussed in this book. Special thanks go to several who have had far greater influence on me than they would ever have guessed. These include Joe Kenney, Frank Retallick, Joe Gallagher, Rod Jones, Ed Kreh, Fred Henning, Lyle Connell, Nate Moore, Ralph Barra, F. X. Brown, and Bernie Hyland.

I also want to thank Jack Martin from the Westinghouse Muncie Plant for his excellent assistance in obtaining the various illustrations used throughout the book. And, finally, I am deeply grateful for the fine typing support I received from Kathy Reynard, Sally Taylor, Cindy Mills, and Cindy Dorste at the Westinghouse Plant.

It is my hope that this book will be helpful to engineers and their managers in understanding and using design assurance techniques. It is a process which contributes to the creation of conditions for excellence in engineering. For in the long run, it will take excellence, and nothing less, to meet the challenges of the complex and ever-changing world we live in.

John A. Burgess

Contents

Preface   v

1. Introduction to Design Assurance   1
   1.1 The Evolution of Controlling Product Quality   1
   1.2 The Concept of Design Assurance   2
   1.3 The Role of Engineering Management   5
   1.4 Relationships with Other Disciplines   8
   1.5 Starting a Design Assurance Program   10
   1.6 Summary   15

2. Design Requirements   16
   2.1 Introduction   16
   2.2 External Sources of Requirements   18
   2.3 Internal Sources   20
   2.4 Functional Analysis   22
   2.5 Translation of Requirements into Designs   27
   2.6 Summary   31

3. Drawing Control   32
   3.1 Introduction   33
   3.2 Types of Drawings   33
   3.3 Drawing Preparation   46
   3.4 Drawing Review and Approval   50
   3.5 Drawing Document Control   53
   3.6 Drawing Revisions   57
   3.7 Summary   62

4. Specification Control   63
   4.1 Introduction   63
   4.2 Types of Specifications   64
   4.3 Specification Preparation   72
   4.4 Specification Review and Approval   76

4.5   Specification Document Control   78
4.6   Specification Revisions   81
4.7   Summary   83

5.   Configuration Control   84
     5.1   Introduction   84
     5.2   Configuration Identification   86
     5.3   Control of Changes   90
     5.4   Configuration Verification   102
     5.5   Summary   103

6.   Design Methods and Analysis   104
     6.1   Introduction   104
     6.2   Design Methods   105
     6.3   Design Analysis   114
     6.4   Design Integration   122
     6.5   Summary   122

7.   Control of Engineering Software   124
     7.1   Introduction   125
     7.2   Software Design   128
     7.3   Coding and Debugging Computer Programs   132
     7.4   Program Verification   138
     7.5   Documentation and Control of Computer Programs   141
     7.6   Summary   149

8.   Product Testing   151
     8.1   Introduction   151
     8.2   Initiating the Test   155
     8.3   Performing the Test   159
     8.4   Reporting Results   161
     8.5   Summary   165

9.   Design Reviews   166
     9.1   Introduction   166
     9.2   Formal Design Reviews   167
     9.3   Requirements Review   179
     9.4   Design Verification Reviews   179
     9.5   Informal Design Reviews   181
     9.6   An Integrated Approach   181
     9.7   Summary   182

10. Statistical Tools for Design Assurance   183
     10.1   Introduction   183
     10.2   Frequency Distributions   184
     10.3   The Normal Curve   188
     10.4   Process Capability   192

10.5   Statistical Tolerancing   195
10.6   Summary   197

11. Control of Nonconformances   198
11.1   Principles of Control   198
11.2   Nonconformance Reporting   201
11.3   Disposition of Nonconformances   203
11.4   Closeout and Feedback   204
11.5   Summary   205

12. Engineering Records   207
12.1   Introduction   207
12.2   Methods for Indexing   208
12.3   Filing Practices   211
12.4   Records Retention   218
12.5   Summary   220

13. Supporting Documentation   221
13.1   Introduction   221
13.2   Operation and Maintenance Instructions   222
13.3   Replacement Parts Lists   228
13.4   Design Control Measures   231
13.5   Summary   232

14. Reliability Improvement   233
14.1   Introduction   233
14.2   Reliability Reporting System   235
14.3   Measuring and Analyzing Product Reliability   238
14.4   Reliability Design Tools   244
14.5   Reliability Testing   254
14.6   Reliability Assessment   256
14.7   Summary   257

15. Auditing the Engineering Process   259
15.1   Introduction   260
15.2   Fundamentals of Systems Auditing   260
15.3   Areas for Investigation   267
15.4   Ethics of Auditor Conduct   270
15.5   Improving the System   272
15.6   Summary   274

16. Design Assurance in the Future   275
16.1   Engineering Management Responsibilities   276
16.2   Drawings   276
16.3   Specifications   277
16.4   Configuration Control   277
16.5   Design Methods and Analysis   277
16.6   Engineering Software   278

16.7   Product Testing   278
16.8   Design Reviews   278
16.9   Statistical Tools   278
16.10   Control of Nonconformances   279
16.11   Engineering Records and Supporting Documentation   279
16.12   Reliability Improvements   279
16.13   Conclusions   280

Appendixes
    Appendix 1: Equipment Specification Contents   283
    Appendix 2: Sample Design Review Checklist   287
    Appendix 3: Guidelines for Auditing the Engineering
                 Function   289

Selected Readings   291

Index   297

DESIGN ASSURANCE
FOR ENGINEERS AND MANAGERS

1

Introduction to Design Assurance

1.1 The Evolution of Controlling Product Quality   1

1.2 The Concept of Design Assurance   2

1.3 The Role of Engineering Management   5

     1.3.1 Setting Policy   5
     1.3.2 Giving Direction   7
     1.3.3 Provide for Training   7
     1.3.4 Monitor and Evaluate Results   7
     1.3.5 Maintain the Pursuit of Excellence   7

1.4 Relationships with Other Disciplines   8

     1.4.1 Marketing and Engineering   8
     1.4.2 Purchasing and Engineering   9
     1.4.3 Manufacturing and Engineering   9
     1.4.4 Quality Assurance and Engineering   10
     1.4.5 Accounting and Engineering   10

1.5 Starting a Design Assurance Program   10

     1.5.1 Self-Examination   11
     1.5.2 Think Big, Start Small   14
     1.5.3 Follow-On Activities   15

1.6 Summary   15

1.1 THE EVOLUTION OF CONTROLLING PRODUCT QUALITY

For many years the control of product quality was considered a factory responsibility. It was largely a process of inspection—try to find the bad ones and separate them from the good ones. However, as mass production expanded, people began to realize the costs and inefficiencies of this approach were very high.

This awareness gave birth to a new and broader outlook on the management of product quality. Shortly after World War II various specialists in the quality control field developed the concept of Total Quality Control. It is based on the premise that quality must be considered and factored into every aspect of the product—from design through production and delivery. Under this approach, all facets of a business are integrated to achieve the desired levels of quality.

One of the major contributions of the Total Quality Concept is the recognition of the importance of preventing defects from happening. The old adage, "An ounce of prevention is worth a pound of cure," is especially applicable to modern industry. Detecting non-conforming or defective products is a difficult and expensive process. Even then, it is often only marginally successful.

A far better approach is to prevent the defects from occurring initially. This approach includes having the right requirements and the proper design as well as seeing the product is manufactured correctly.

In addition to the Total Quality Control concept, other means for enhancing the quality and effectiveness of product design have been developed over the past several decades. One notable effort has been cultivated by the American Society of Mechanical Engineers. In the 1920s and 1930s each boiler manufacturer designed and built his product according to their own methods and practices. Periodically, this resulted in boiler failures which caused extensive damage and loss of life. Finally a small group of engineers affiliated with ASME took it upon themselves to develop a standard set of rules and practices for all manufacturers to use. This led to safer designs, and the approach has been very successful through the years in dramatically reducing boiler and pressure vessel failures.

Many techniques are now being applied in industry to prevent defects from occurring in design and production. This focus on prevention has been a major contributing factor in the evolution of design assurance.

1.2 THE CONCEPT OF DESIGN ASSURANCE

Design Assurance, also referred to as Design Control, is a relatively new term in our industrial vocabulary. It has developed slowly from Drawing Control and expanded to cover all facets of product engineering activities.

For the purposes of this book, Design Assurance is defined as:

> . . . those planned and systematic actions taken to provide confidence that the product design will satisfy the requirements of its intended use.

It is a response to the rapid advances in technology, increased complexity and growing sophistication in industry. No longer is it acceptable to rely on the "business-as-usual" methods and practices of years gone by.

The Design Assurance process is an outgrowth of management methods developed and used on military, aerospace and nuclear projects. Various quality systems were developed in the 1960s and 1970s to manage the process of achieving the proper levels of quality. Examples of these include MIL-Q-9858A, NASA NHB 5300.3, AEC-10CFR50 Appendix B, and ASME/ANSI NQA-1. Each of these program requirements documents contained provisions for controlling engineering activities.

Understandably, the application of special quality system requirements on engineering activities came as quite a surprise to many people. Always before, quality control focused only on the factory. Yet under closer examination, experienced engineers and technical mangers will recognize that most of the elements of Design Assurance are simply good engineering practices.

Why then all the fanfare about some new discipline called Design Assurance? The best way to answer that question is with some real-life examples.

During the construction of a large nuclear powerplant, a major earthquake fault was discovered a few miles away from the powerplant site. After much investigation and analysis, it was concluded that additional seismic supports were required. The owner of the powerplant hired a competent engineering firm to do the redesign. The owner furnished a sketch of the plant arrangement which was to be used in the design project. However the sketch was "very sketchy." It did not contain all of the information that was required and, in fact, was based on some outdated drawings. However, the design group believed they had the correct information that was needed. As a result, the design work was accomplished, but it was not usable because it was based on incorrect input.

After the design work was finally corrected, the owner asked that the new design be applied to the adjacent, companion unit. However, the owner failed to tell that the companion unit was not identical to the first, but was a mirror image of it. Again the results were not usable until further corrections were made.

A medium-sized manufacturing firm produced a line of single and multi-stage pumps and enjoyed a strong market position in their industry. To the management's delight, they received a very large order for their pumps. However, the application required some development and modification of the standard designs. During the development phase, the engineers made a series of changes to the internal configuration. However, when the first new pumps went into production,

several of the impellers cracked and broke during acceptance testing.
After a detailed investigation, it was found that the structural analysis
was performed on the next-to-last revision, but not on the final change.
The last modifications caused the structural limits to be exceeded, but
no one realized the shortcoming until the pumps failed mechanically.

Another firm entered the microprocessor-controlled equipment area
after many years of success as a builder of hydromechanically controlled
equipment. Although the design and development efforts proceeded
cautiously, the overall knowledge and skill of the design force was
limited in the field of electronic controls. Considerable time and money
was wasted as a result of inadequate technical review of the design.
Many errors and oversights were made on matters which otherwise would
have been found and corrected by qualified electronics designers.
Finally, the management realized the shortcomings and obtained the necessary expertise for the program, and just in time to prevent a total
disaster.

Now, to answer the earlier question: Why all the fanfare about
Design Assurance? It is in recognition of the major impact that design
has on the long-range well-being of a project, a product, or a producer.
Today's consumers are much more aware of, and concerned about, the
quality of the products they buy and use. Management now is realizing
it is very difficult and costly to try to make a good product out of a
poor design.

Errors or shortcomings in design are often hidden until the design
becomes hardware. Even then, design problems may not show up until
many units have been produced and are in operation. When a problem
is discovered at this stage, it may be both expensive and embarrassing
to correct. Yet, the penalties of the marketplace, and more recently of
the courts, for failing to correct a problem, may be even worse.

Design Assurance is a response to these conditions. It is another
management tool for running the business in an efficient and effective
manner. And it should be management's responsibility. Several years
ago, Dr. Juran, a prominent authority in the field of quality, found in
his studies of industrial problems, that the workers cause or can correct
only about 20% of the quality problems. The remaining 80% of the
problems are caused by management, or are within their ability to influence or control.

Design Assurance is also a direct response to the need for defect
prevention. Instead of allowing the product to get into production and
then seeing if it will work, it is an approach for anticipating what can
go wrong and preventing it from occurring. This also includes examining past designs and their problems and taking action for preventing
recurrence of the old problems in the new designs. Do it right the
first time is the theme of Design Assurance.

Design Assurance covers many traditional areas of engineering,
such as drawing and specification control, but it also includes broader
facets, such as, a disciplined approach to design decisions, verification of design adequacy, control of engineering software, proper

integration of interfacing designs and design activities, and the interaction of product feedback with the design process.

As the state of the art of technology advances and the complexity
of products and systems increases, it is no longer possible to rely on
the handbooks of old and the seat-of-the-pants method of engineering
a product.

Greater emphasis must be placed on a more disciplined, but manageable, approach to engineering than has been used in many industries. Yet, the process must not be so rigid or overbearing that it
stifles creativity or ingenuity. Also it must be responsive to cost and
time demands to be competitive and to the resource limitations of the
firm. Obviously, these are challenging criteria to be met by engineers
and their managers. And meet it they must.

The subsequent portions of this book are intended to help those
same persons with suggestions and recommendations on how to accomplish that feat.

1.3 THE ROLE OF ENGINEERING MANAGEMENT

Engineering management must take the lead in the introduction, development, implementation and maintenance of the Design Assurance program. It is not something that naturally grows from the bottom up.
It requires imagination, leadership and fortitude to make it happen and
do it without mutiny or malpractice.

Engineering management sets the tone for the engineering organization in its attitude and actions on matters involving engineering
excellence. How does the engineering manager respond when faced
with a surprise problem? A conflict between a marketing requirement
and a preferred design practice? A choice between cost and quality?
How high is his tolerance for errors? Decisions and actions on these
and similar situations influence the engineering organization's attitude
toward the pursuit of excellence in their work. Each engineering manager needs to reflect on their own actions. Do your actions match
your words?

1.3.1 Setting Policy

Establishing policies for assuring the quality of design is Engineering
management's responsibility. These policies need to be written for all
to see and use in the day-to-day decision-making process. A policy
statement need not be written in the poet's prose or in lengthy text.
In fact, the most effective policies tend to be brief and to the point.
One such policy statement is shown in Figure 1.1. It is for a company
deeply involved in high technology and major technical systems. In
contrast, the policy statement presented in Figure 1.2 is for a relatively small firm which operates in a specialty product niche. This company works with relatively conventional and straightforward products
and problems.

DESIGN ASSURANCE POLICY

It is the policy of the Engineering Department to strive for excellence in all of its efforts related to design and development of products for our customers. Engineering management is dedicated to employ only persons of vision, skill and integrity and to provide them with the tools, facilities and resources to perform their work with precision and accuracy. Engineered products shall be designed in strict accordance with the applicable codes, standards and specifications. Each new design shall be thoroughly evaluated and proven, using the latest methods and practices available, prior to releasing it for operational use.

In the event that defects or discrepancies are identified, appropriate corrective actions shall be taken promptly to correct the problem, and measures shall be implemented to prevent future recurrence of the conditions adverse to quality.

FIGURE 1.1 Sample policy statement for a high-technology company.

DESIGN POLICY

The Engineering Department shall exercise care and concern in the design of the company's products. Appropriate methods shall be used to assure the designs meet the performance, reliability, cost and safety requirements while providing products our customers consider to be of value to them. Each person is expected to perform their work accurately in accordance with the proper methods and to report to management any instances where errors or discrepancies are found.

FIGURE 1.2 Sample policy statement of a company with simple products.

Such policy statements frame the concepts of design assurance as they are to be applied by those particular companies. The policy becomes the foundation for the building of the design assurance structure. And, of course, it serves as an important yardstick; it continually raises the question: How are we measuring up?

I.3.2 Giving Direction

But policy alone is not enough. It needs to be converted into direction and action. Engineering management must provide the leadership to get the program functioning. They must see to it that the necessary resources are applied to develop applicable methods and procedures. For without these, the program is simply a figment of the imagination.

The methods and procedures must address the details of how the work is to be performed. Answers to those basic questions of: What? When? Where? Why? and Who? need to be answered according to local needs. It takes time, thought and a bit of trial-and-error to come up with workable procedures. So don't be discouraged if the first cut seems awkward or only partially successful. Engineers rely heavily on the iterative process to perfect their designs. The same process applies equally well to the development of their administrative procedures.

1.3.3 Provide for Training

Once the methods are developed, management must make provisions for training the affected persons and groups to use the new methods. It does not have to be an elaborate program. Simply concentrate on two key points: (1) How you want the work performed? and (2) Why it is important to do it this way? Answers to both questions are necessary to achieve the desired results.

1.3.4 Monitor and Evaluate Results

Engineering management also must take the time periodically to monitor the progress and evaluate the results. Is the work really being done in the desired manner? Are we getting the expected results? Is it being accomplished in a timely and cost-effective manner? These are questions that management needs to ask and to get answered.

Later in the book, guidelines and instructions for auditing the engineering activities are described (see Chapter 15). This approach is an effective means of gathering data for the evaluation process.

1.3.5 Maintain the Pursuit of Excellence

To get the maximum benefit from a design assurance program, it must become an integral way of life. It can not be like a fancy coat you put on for show on special occasions and then put back in the closet

after the event is over. Design assurance needs to be an active and
on-going part of doing business. It is a tool for all to use in the
pursuit of excellence in engineering. Excellence is used in the sense
of value—a product or service which the customer considers is
particularly useful and at a price which the buyer finds especially
attractive. Words such as "dependable," "reliable," "trouble-free,"
"the best," are often used by customers in describing their personal
understanding of value and excellence.

As part of this effort for achieving excellence, Engineering management must be actively involved in preventing defects. This requires an on-going commitment for improving the quality of design
through corrective actions. In the heat of battle it is common to cover
an injury with a band-aid. But in the long run, major surgery may
be required to repair the damage or cure the disease. It takes both
courage and commitment to eliminate causes and not simply treat symptoms. Consequently, engineering management must face those decisions squarely. They are expected to take the steps needed to prevent
design problems from inflicting serious wounds on the company. Personal pride, NIH factors, inventor's attachment and "ivory-tower"
isolation are stumbling blocks along the path. Regardless of these
difficulties, enlightened management is obligated to press ahead to
achieve quality in design. The pursuit of excellence is a never-ending
journey.

1.4 RELATIONSHIPS WITH OTHER DISCIPLINES

Engineering typically enjoys an element of respect from other departments due to the specialty knowledge and skills and from its traditional role as the decision maker on technical matters. However, some
engineering departments also are viewed with a certain element of dis-
respect. This may be due to past actions which seem to say to others
that Engineering is "too good to get their hands dirty in the daily
problems." Unfortunately, respect can only be earned and not
legislated.

Engineering's role is typically such that they influence nearly all
other groups and functions. Thus, Engineering cannot be an island
unto itself. And for a company to reap the benefits of profitability,
high productivity, customer credibility, and workforce pride that a
Total Quality Commitment can offer, Engineering must be an active and
responsive member of the team. Again, the leadership and direction
must come from Engineering management.

Let's look at several of the relationships Engineering needs to
cultivate within the company as part of its development of a design
assurance program.

1.4.1 Marketing and Engineering

To obtain the maximum favorable impact from the design assurance

program, Engineering needs to work closely with Marketing. Details
of customer requirements, market trends, competitors' actions and
products, and feedback on the company's product performance are
important inputs to the engineering process. It is imperative that the
design engineer receives the correct customer requirements and is advised of market developments and field problems in a timely manner.

Conversely, it is equally important for Engineering to keep Marketing advised of promising new developments, the need for design
changes, the introduction of design improvements, etc. Periodic
meetings between the design and sales groups are often helpful in this
regard.

It takes this kind of teamwork to be an effective competitor. Also
a good Engineering—Marketing interface is crucial for enhancing the
quality of design. Remember, it is not enough to do the design right;
it must be the right design for the market. And, unfortunately, there
are a lot of well-designed "buggy whips" to prove the point.

1.4.2 Purchasing and Engineering

It is common to find that 30-50% of a product's cost comes from pur-
chased material and parts. This obviously has a major influence on
the finished product. To adequately control this important element,
Purchasing and Engineering need to work together closely in many
areas. This includes: the selection and qualification of suppliers, a
thorough and accurate definition of what is wanted, the resolution of
supplier's questions and problems, and evaluation of the actual performance of the parts and materials the supplier provides. Engineering must be willing to investigate requests for changes, resolve am-
biguities, and occasionally relax some of the requirements. Purchasing
and Engineering can make major contributions in cost control, product
performance and timely resolution of problems.

1.4.3 Manufacturing and Engineering

Although there is a long-standing tradition of conflict between design
and production, that is a condition that must be changed to be successful in today's markets. Otherwise, the company will fall short in
its efforts to enhance the quality of its designs and products.

In many instances the interface between Engineering and Manufacturing is much like the interface between Engineering and Purchasing. The Manufacturing Department and outside suppliers have needs
from Engineering that are quite similar, e.g., clear and complete
product definition, control of changes, realistic tolerances, timely
resolution of questions and problems, etc.

From Engineering's standpoint, they have every right to expect
Manufacturing's compliance routinely with realistic design requirements.

Many of the basic elements of Design Assurance covered in subsequent chapters, such as design and specification control, control of

changes, and resolution of non-conformances, are tightly interwoven
with the Engineering—Manufacturing interface. Each of these interactions provide an opportunity for teamwork to get the best possible
results. It's tough enough to fight the competition; don't make it
tougher by having a running feud between Manufacturing and
Engineering.

1.4.4 Quality Assurance and Engineering

The interface between Quality Assurance and Engineering is a natural
one. Both are interested in conformance with the requirements and
with the resolution of quality problems. However, Engineering frequently fails to recognize QA's need for information and thorough
understanding of design intent. Also design engineering underestimates
the contribution that QA's data can make in describing how a product
or process is really working. The advances in technology make it necessary for Engineering and Quality Assurance to work hand-in-hand
throughout the product life cycle, from conceptual design through
production and operation. It takes this kind of interaction to gather
and apply real-world data to the company's designs.

1.4.5 Accounting and Engineering

One interface that is frequently neglected is the one between Accounting (or Controller) and Engineering. Granted, many of the tasks that
the financial control group performs have little effect on the design of
the product. However, those aspects of product cost accounting
should and do play an important role.

For effective design control, Engineering needs to know and appreciate how its designs affect the cost of the product and what the
impact of new designs is on factory costs, inventories, tooling, etc.

The cost data can provide useful insights and feedback to the
Engineering organization, and these should not be overlooked in the
Design Assurance effort. In fact, the increasing growth of interactive design tools now makes it more important than ever to consider
the impact of cost as various designs are considered and evaluated.

1.5 STARTING A DESIGN ASSURANCE PROGRAM

Before making a big deal about starting a new program called Design
Assurance, the management should recognize that some elements probably are already in place. Nearly every engineering organization has
some methods it uses to develop new products, to define what the shop
has to make, and to get information about changes to at least a few
persons who need to know. It is often the degree or extent of the
process that may need to be revised or strengthened.

It is also quite possible that the management and long-time professionals may not be familiar with the term "Design Assurance" or appreciate what it typically involves.

Therefore the starting point should be a review of present practices and a comparison with various elements frequently found in effective Design Assurance programs.

1.5.1 Self-Examination

The Table of Contents and the subsequent chapters of this book present the major elements and recommended practices to apply in Design Assurance programs. Obviously, it is desirable to read and consider these elements carefully in preparation for the review of the present practices used locally.

Table 1.1 is an abbreviated compilation of key Design Assurance program elements. It is arranged in a manner to suggest a wide range of possible coverage. The columns on the left represent little or no control, while the column on the extreme right represents a comprehensive, integrated approach to Design Assurance. Upon examination of the local program, the person or persons responsible for the review may find one or more elements are performed in an acceptable manner. However, other elements may be totally lacking or need strengthening.

When reviewing existing practices, the evaluator needs to seek answers to several key questions. These questions are:

1. What are we doing now?
2. How well does it work?
3. What else needs to be done?

The questions should not be answered lightly. Some investigation of actual practices, results, recurring problems, etc., should be made. Don't simply assume everything is done in the manner management would like for it to be done. Take a first-hand look. Check some of the indicators, such as, the number of changes being processed, volume of shop requests for waivers or revisions, product failure rates, customer complaints, drafting hours per drawing produced, accounting charges for engineering errors, etc.

Be reasonably methodical in the investigation. Look at various elements and draw conclusions on what you find. Include drawings, specifications, calculations, records, development tests and design changes in your evaluation. Which areas appear to be satisfactory, at least for the time being? Which areas appear to be sources of problems? Which areas aren't being addressed at all but probably should be? Note which ones seem to be satisfactory and which ones appear weak. These conclusions should then serve as the key inputs for deciding where to start.

TABLE 1.1 The Evolution of Design Assurance

	No program $\longrightarrow$				Fully integrated program
Design requirements:	The designer figures out what's needed.	The salesman sends spec sheets.	Occasional meetings between Engineering and Sales managers to discuss new developments/marketplace happenings.	Engineer and salesman work together on bid/contract reviews.	Marketing–Engineering team interactive review process. Timely feedback.
Drawing control:	The designer makes sketches and notes for the shop.	Drafting prepares drawings per designer's instructions.	Engineers' supervisor reviews selected drawings.	A few technical specialist's review and sign selected drawings.	On-going series of design reviews. Multidisciplined signoff.
Control of changes:	Verbal instructions from engineer.	Drawings/specs marked up by engineer.	Engineer issues written change notice.	Engineering management reviews/approves written changes.	Technical specialists review written changes to evaluate impact and approve as appropriate.
Engineering calculations:	Parts sized by handbook and seat-of-the-pants engineering.	Engineer performs written calculations based on personal methods.	Engineering calculations for selected items reviewed by supervisor or lead engineer.	Standard methods based on analysis and test data developed and tailored per product needs.	Latest state-of-art interactive design systems in place with necessary support specialists.

Engineering tests:	None. Let the user test it for us.	Engineer does some experimenting in the shop.	One engineering model tested by outside lab.	Test of full-scale hardware. In-house test facilities. Test engineers on staff.	Planned and funded development test programs New products proven by factory and field tests.
Control of nonconformances:	There aren't any, everything goes.	The factory decides if it's good enough.	The engineer is consulted if it looks bad, but we need it.	The engineer decides if it is good enough.	Material Review Board of technical specialists dispositions non-conformances.
Engineering records:	Don't need any. Keep it in our heads.	The engineer keeps a folder of notes and things somewhere.	The department clerk collects and files the drawings and records.	Records procedures and centralized files used to control and retain records.	Engineering records are controlled as integral part of company records management program.
Management evaluation:	The engineering department does its own thing—and nobody notices.	When a problem occurs, management wants to know who goofed.	When a problem occurs, management wants to know what went wrong.	Management hires or designates a knowledgeable person to review the engineering activities every few years.	A planned program of internal audits by engineering and other departments conducted regularly. On-going program of refinement and improvement of management processes.

It's a good idea to record the conclusions and the basis for them
for each of the areas examined. Use these notes in the planning process, and save them for future reference as the program gets underway.
It's easy to forget why certain actions were considered necessary as
the plan is implemented and often modified along the way.

1.5.2 Think Big, Start Small

Look down the road to the future. How far do you want to go with this?
How broad a program do you eventually want to have? And how soon do
you want to get there? Generally, a good approach is to proceed initially on a small scale. Select one or two areas which appear to be logical
opportunities for improvement. Concentrate on these and learn by doing. Don't dilute your effort by trying to fix many things all at once.
That approach can be both disruptive and ineffective.

When planning the development of new Design Assurance methods,
recognize that each organization must tailor its practices to its own local
situation. The methods chosen should reflect the size and personality
of the company, the complexity of the product, the competitive situation.
the skills of the person who will be affected by it, and any other factors
of significance to the particular firm. Some companies are very rigid,
others quite flexible. Some are very formal, and others informal.
Nevertheless, to be effective, the Design Assurance measures chosen
must be applied with a degree of discipline and integrity. Seek methods
you can live with and then live with them faithfully. That doesn't mean
you follow a rigid course into obvious chaos or disaster. But don't
arbitrarily bend the rules at every turn of events. Give the process a
chance to work for you.

Much has been written in the management literature about the best
ways for introducing change. And most of it applies to introducing
new methods of operation, like Design Assurance. Take time to plan,
and do your homework. Communicate your plans and intentions and
involve the people affected. Get more than one input on how to do it.
Listen and be responsive to questions and comments from the people
that will be applying the new methods. And put a knowlegeable leader
in charge of making it happen.

As a new method is developed, take time to train the persons that
must use it. It's not enough to know what is to be done and how it
should be accomplished. People nowadays, especially professional and
administrative people, want to know more. They want to know why it
is to be done this way and what is significant about this particular
method. In the long run, it is very helpful and typically gives better
results. When people understand the process, there is a much greater
probability that the intent will be achieved.

Last but not least, management must then give the persons assigned
a reasonable time to develop and debug the process so it becomes an effective method of operation.

1.5.3 Follow-On Activities

After the first one or two elements have been implemented successfully, then select the next couple of elements to improve. Use the same care and approach on the second phase as was used successfully on the first. Don't assume it will automatically happen now. In this manner, ease into the program, and it will provide benefits with the minimum of disruption.

As the implementation of the various program elements proceeds, be alert to the interaction of the various elements. For instance, the new system for controlling changes to drawings might also work well for controlling changes to specifications or instruction manuals. The process for checking hand calculations might also apply to verifying new or revised computer programs. Methods for filing and retaining design records might also work for test records. This is often a synergystic process that leads to further refinements and efficiencies.

As the various elements of the Design Assurance program are introduced, management should monitor the process and look critically at the results. Management has the right and obligation to expect new methods to give improved performance and do it in a cost-effective manner. Anything less should be re-examined and modified or eliminated.

Management evaluation is important for several different reasons. First, to find out if the new method is actually being used. Second, to determine if it is giving the intended results. Third, is it efficient to do it this way in actual practice? Are there any wrinkles that need to be ironed out? And, finally, now that people are using it, do they have suggestions for improving on it?

The evaluation process also conveys a message of commitment on management's part to make it work. Again, management's actions speak louder than its policies or pronouncements.

1.6 SUMMARY

Design Assurance is an important part of a company's broad commitment to quality. It is the process of applying sound engineering practices to all engineering tasks. Furthermore, it reflects management's commitment to excellence in its products. Engineering management must take the lead in determining the scope and direction of the Design Assurance program. They must also provide the necessary push to make it happen. From there on, the program should stand the test of time. It should be expected to pay for itself in the long run through the prevention of problems and in satisfying real market needs. Anything less should not be tolerated.

2

Design Requirements

2.1 Introduction   16

2.2 External Sources of Requirements   18

    2.2.1 Customer Specifications   18
    2.2.2 Industry Standards   18
    2.2.3 Product Feedback   19

2.3 Internal Sources   20

    2.3.1 Marketing/Sales Requirements   21
    2.3.2 Product Development Requirements   21

2.4 Functional Analysis   22

2.5 Translation of Requirements into Design   27

    2.5.1 Design Checklists   28
    2.5.2 Requirements Allocation Sheets   29
    2.5.3 Incorporation into Drawings/Specifications   29

2.6 Summary   31

2.1 INTRODUCTION

The starting point for all design work is with the definition of the design requirements. Although this seems elementary, it is surprising how little visibility this phase of design often receives. Design engineers routinely consider and follow requirements from many sources, but there is seldom a record which shows what the requirements are, where they originated, or why they are important. This limited visibility makes it difficult to verify if the necessary requirements have been considered and whether or not there are excesses or omissions that should be resolved.

Requirements for a new or modified design can come from several sources. Some are internal to the company, and some are from the outside. It is the job of the designer to gather the various inputs, determine which ones are important to the design effort, and then convert these into the drawings and specifications for the product as shown in Figure 2.1.

The time and effort invested in the definition of the requirements has a high rate of return in the avoidance of problems later in production and operation. It is a simple matter to change a drawing early in the design phase, but it can be very costly to change the product after it is being manufactured.

This chapter describes the various sources of design requirements that the designer should investigate and consider. A special means for determining design requirements, called functional analysis, is explained, and examples of how to use it are included. Also several tools which are useful to the designer in recording, compiling and cataloging design requirements for use in the design process are presented. These approaches are helpful in organizing the information and making it easier for the designer to use it effectively.

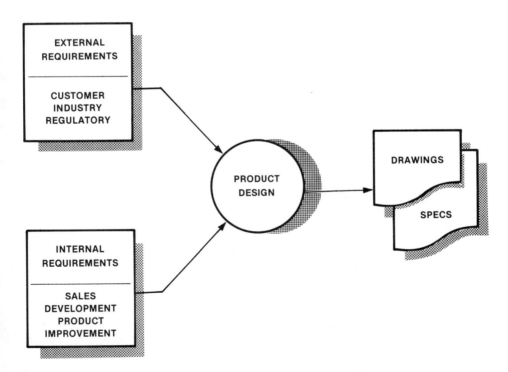

FIGURE 2.1 The cycle of design.

2.2 EXTERNAL SOURCES OF REQUIREMENTS

When designing a new product or modifying an existing one, requirements from external sources should be among the first things
considered. These sources include requirements specifically stated
by the customer, recent changes in industry standards, and, of
course, feedback from customers, dealers and company service calls.
Each of the areas will be examined further in this section.

2.2.1 Customer Specifications

Many industries provide products which are tailored to a customer's
specified needs, such as, material and equipment for industrial applications. The purchasers of industrial equipment tend to be more
knowledgeable about technical features and more demanding in terms
of specific requirements than the consumers in the general public.
Therefore, it is quite likely that an industrial purchaser will prepare
and issue a specification, defining what is wanted in considerable detail. The specification is then distributed, and potential suppliers
are invited to submit a bid for the work.

Experience shows that some suppliers pay very little attention to
a buyer's specification. Often, the result is their bid is not accepted.
Granted, many factors go into a purchaser's decision such as price,
delivery, demonstrated capabilities, perceived value, etc. Nevertheless, a supplier who fails to make an offering that addresses the requirements of a customer's specification generally is increasing the
risk of being an unsuccessful bidder.

From a design assurance standpoint, the customer's specification
should be used as a checklist in the design process. Although some
requirements may be unclear, or even contradictory to the designer,
each requirement in the specification should be carefully considered
and factored into the design to the maximum extent possible. Where a
requirement is confusing or the intent is uncertain, it is a wise practice
to request clarification from the customer. In those cases where the
requirements are at odds with the company's or designer's intent, it is
up to the designer to make such differences known to his management.
Decisions to deviate from a customer's specified requirements should
not be made arbitrarily, since such actions may have a big impact on
whether or not your bid will be a winner.

2.2.2 Industry Standards

Another important source of requirements that the design engineer
needs to consider are applicable industry standards. Some standards
are very rigid and literally carry the force of law in many states or
municipalities. One such standard is the ASME Boiler Code for various
types of pressure vessels. Other standards may be voluntary, but

are important because of the need to be compatible with accepted industry practices. Compliance with some standards may also be necessary to achieve interchangeability with existing, competing items
which are already in use.

Industry standards are usually prepared and maintained by trade
associations or technical societies, like the National Electrical Manufacturer's Association (NEMA), American Welding Society (AWS),
American Society for Quality Control (ASQC), and the American National Standards Institute (ANSI), to name a few.

Membership on standards writing groups is normally comprised of
volunteers from manufacturers, consultants, government agencies,
etc., who are actively engaged in the industry affected by the standards. Since the resulting standards often have a significant impact on
the particular industry, design engineering groups need to follow the
standards development process and be knowledgeable of how new or
revised standards will impact their designs.

Another type of standard also is becoming important in many industries today. It is in the form of a regulation issued by a state
or federal agency. Although industry has some influence and effect
on the regulatory process, it is certainly much less than what can be,
and is, exerted in the voluntary standards activities. A government
regulation often tends to have the force of law. It may well have to
be accommodated in the design effort and compliance verified through
the design assurance process.

2.2.3 Product Feedback

A wise, old engineer once said, "Listen to the product when it is
speaking." Truly, this is sound advice (no pun intended) but so
often ignored. Products speak to us in many ways in the factory and
in the field. They speak through their performance; both good and
bad. They speak through their users and their complaints. They
speak through returns from dealers and orders for replacement parts.
They speak through the results from tests and problems encountered
during manufacturing, installation and checkout.

These, and other channels, provide valuable information to the
design engineers. Yet, many do not bother to look at the data, until
a calamity occurs.

On the other hand, some firms have developed the use of product
feedback to a fine art. Their designers are closely attuned to problems, or lack of problems, and use this knowledge routinely in their
design work. It is equally important to know what is working well and
avoiding changes which could have an adverse effect as it is to know
of a problem that needs to be resolved.

However, it takes effort and skill to obtain reliable data about how
the product is performing. No news isn't necessarily good news. The

TABLE 2.1   Sources of Product Feedback

Prototype/Development test results

Final test results

Inspection reports, nonconformance reports and rejection statistics

Customer returns

Customer complaints

Dealer complaints

Analysis of factory or field failures

Shop complaints and requests for changes or waivers

Audits/special checks made on a sample of products which are ready
    for shipment to the customers

Orders for repair or replacement parts

Industry reports on problems or performance/reliability statistics

Information exchanges at technical conferences or trade shows

Surveys of customers by dealers or sales personnel

Service reports from dealers or authorized service personnel

Meetings with customers or users

---

design organization needs to be monitoring a product's vital signs in
an on-going manner to avoid being "blindsided" by a major problem.
    There are several different sources of product feedback which
are, or can be, available to nearly all engineering groups. Table 2.1
lists numerous avenues for obtaining information about a product's
state of health.
    By monitoring several different sources of feedback information,
a design engineer can keep his finger on the pulse of the product.
No single source alone is sufficiently reliable to depend upon exclusively. However, by collecting and periodically reviewing the data from
the various sources, the designer can frequently recognize when the
product requirements need to be changed. This is an area that should
not be overlooked in the establishment and operation of a design assurance program.

2.3 INTERNAL SOURCES

Although there may be several internal sources of design requirements,
two are especially important. These are the marketing or sales

department requirements and the product development requirements. Both of these interact and are important to the ultimate success of the product in the marketplace. Further discussion of these and the implications to design assurance are discussed below.

2.3.1 Marketing/Sales Requirements

The Marketing or Sales Department of a company frequently specify requirements to the Engineering organization. These requirements may be to satisfy an important customer or to achieve or maintain a niche in the marketplace. Some of the inputs will be feedback about the firm's product and competitor's products from customers or dealers as described in Section 2.2.3. Other inputs will be Marketing's perceptions of what design features or product characteristics must be provided to be competitive.

In any case, the design engineer must recognize the importance of incorporating the Marketing inputs into the design process. Provisions should also be made to verify that these requirements have, in fact, been included and properly translated into the product drawings and specifications.

2.3.2 Product Development Requirements

Like Marketing, Engineering may have a number of special requirements which must be included in the design. These requirements generally come from the product development efforts conducted in the engineering laboratories and test facilities. Such efforts are intended to lead to new or improved products or cost improvements for existing product lines.

Product development requirements present some special challenges to the design engineer. It is quite common in development work to have only limited time or funds to explore new ideas and approaches. Thus, when a new development looks promising, the design requirements probably are not sharply defined, and some of the requirements may be "soft," i.e., not firmly established or limits are uncertain.

Engineering management plays an important role in such cases. The management needs to provide the leadership in the examination of the results and trends in the product development work. They should participate in the decisions, regarding which parameters and features should be incorporated in the designs.

From a design assurance standpoint, it is necessary to identify those new features or assumptions that need additional investigation later in the design verification stages. Failure to recognize and appreciate the added level of risk or uncertainty in a new design feature can lead to problems later in manufacturing or in service. Tracking of "soft" requirements and follow-through to solidify them is where the design assurance process can be most helpful.

2.4 FUNCTIONAL ANALYSIS

Occasionally, an engineering organization faces the challenge of designing a product which is truly new and unfamiliar to the company. When
this occurs, it may not be possible to draw heavily on prior products
for guidance. Thus, a different approach is needed.

Some years ago, a systems engineering technique, known as functional analysis, was developed for just such situations. It starts with
a statement of the overall objective or mission and then expands this
into various functions which must be accomplished to achieve the
objective.

The technique is sufficiently general in nature that it can be applied to a wide variety of problems. Although the method imposes a
certain amount of discipline in its structured approach, the process
reduces the chances of overlooking essential requirements, especially
those that may otherwise be taken for granted and forgotten to be
included.

The first step in the process is to define the objective clearly and
succinctly in writing. Reducing it to writing frequently reveals voids
in our understanding and shows where further knowledge or thinking
is required. A lot of thought should go into the answer of the question: What is the overall objective for the design? Since the analysis
that follows is built on this answer, it is important that it represents
the true objective. It isn't enough to do a good job. You must do
the right job.

To bring the task into sharp focus, write the objective in one or
two sentences or, at the most, one paragraph. If you can't summarize
it in this manner, more thought is needed.

For example, assume the task is to design a new hand tool. The
Sales Department has identified a large potential market among home
owners for a portable power tool for making holes from 1/2-inch to
2-inches in diameter. Thus, the object might be stated as:

> Develop a portable hand tool capable of making holes 1/2-inch to
> 2-inches in diameter. The tool shall have self-contained power
> and be designed for the do-it-yourself market.

This objective identifies a few key parameters (portable, self-
contained power, holes 1/2 to 2 inches in diameter) and the end use
(do-it-yourselfers).

The next step is to review the objective statement with your
supervisor or team leader. See if they agree that it represents the
true objective in their minds. If not, modify it until you reach a con-
sensus. As you proceed with the analyses, keep the written statement
of the objective close at hand and refer to it frequently. It will help
keep the project focused on the desired target.

The next step is to identify the major functions the product must
perform. From the statement of objectives, several key functions

can be named. These functions then should be drawn as a simple
block diagram. The diagram will serve as a logic chart and is part of
the structured method. Figure 2.2a illustrates an example of the basic
functions to be performed and is identified as the Top Level or First
Level, Diagram. From this, the basic functions will be expanded as
the analysis proceeds.

Note that each function is listed in a separate rectangle, and the
rectangles are connected by solid lines with the flow of information
progressing from left to right.

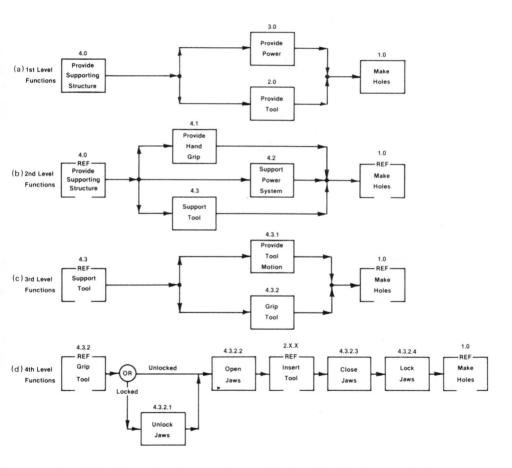

FIGURE 2.2 Functional flow diagram for hand tool. (From J. A.
Burgess, Organizing Design Problems, *Machine Design Magazine*,
Nov. 27, 1969, Penton/IPC, Cleveland, OH, pp. 120-127.)

Systems analysts have developed a number of techniques for pre-paring functional flow diagrams. These provide a standardized method for constructing these visual aids.

The first diagram shows the series-parallel relation of the basic functions. Each block of the diagram is assigned its own decimal number, 1.0, 2.0, 3.0, etc. Number 1.0 is normally assigned to the block furthest to the right, representing the end objective. Beginning with the block adjacent to block 1.0, the other blocks are then num-bered in sequence away from the end point. This number is used for identification and for relating it to subsequent flow diagrams. It is important to recognize that the series-parallel relationship of the blocks takes precedence over the block number sequence. From the first level diagram, the major tasks or features of the product or sys-tem can be identified. As stated earlier, the first level functions are derived from the objectives.

Figure 2.2a illustrates the first level diagram for the hand tool program. From this you can see the four major functions: (1.0) Make Holes, (2.0) Provide Tool, (3.0) Provide Power, and (4.0) Provide Supporting Structure. Although the functions listed in these diagrams may appear to be obvious, greater insight into the problem is needed to define the functions at the lower levels.

This is done by expanding each block of the first level diagram into a separate diagram of its own. More detailed functions are de-scribed at this level, and each new block is assigned a sequential num-ber which relates it to the corresponding first level block. Thus, 4.0 is expanded into 4.1, 4.2, and 4.3 in Figure 2.2b. At the second, and subsequent levels, the assigned block numbers generally progress in the direction of information flow.

In this example, the function of Provide Supporting Structure has been expanded into three functions which are more specific: Provide Hand Grip, Support Power System and Support Tool. For the hand tool design to be effective, it must satisfy each of these functions.

The expansion process is repeated for each new block, and addi-tional levels are developed as shown in Figure 2.2c and 2.2d.

You may ask: How far do you carry this process? And the only answer is: Until you can no longer identify additional functions.

One of the more perplexing tasks in functional analysis is defining each of the functions. Each block represents something that must be accomplished and is described by a verb and a noun, such as, Start Motor, Retain Tool, etc. It must describe a particular action, but in broad terms. Occasionally, the analyst will describe the solution rather than the actual function which must be accomplished. For example, Block 2.0 in Figure 2.2a might have been labelled: Provide Drill. How-ever, this presupposes a design solution and may exclude other ap-proaches to the problem, such as, a mechanical vibrator, rotary cutter, electrical discharge, laser beam, chemical, etc. In many instances, it is acceptable to assume a design approach if you do it knowingly. Nevertheless, be suspicious of functions that can only be accomplished in one way.

This simply points out the need for broad thinking in the functional analysis process. It is a tool which helps the user focus on the total task and understand where the requirements originate. Thus, he can relate the significance of the design requirement to its source. It also helps engineers overcome "tunnel vision" in their work.

Identifying interfaces in functional flow diagrams is another important task. Two different techniques are frequently used to accomplish this. The first of these is the reference block and is shown as a partial block (similar to mathematical brackets). The reference block is used as the title or starting block when expanding a block from a higher level diagram. It is also used to show how a function (block) on one diagram is connected to a block on some other diagram. This is commonly used as the last or right hand block on a functional flow diagram, although it may also appear other places in the diagram. These types of interfaces are shown in Figure 2.2.

The other method of showing interfaces is the use of the dashed line blocks. Some people use dashed lines to show interfaces with functions which are performed external to the originating organization; for example, functions performed by the customer, certifying agency, etc. Others limit the use of dashed lines to delineate tentative functions, thus flagging the function as requiring additional thought.

Another important feature of the functional flow diagrams is the use of logic gates. The "gate" technique provides a means of combining several actions or differentiating among alternate actions. The AND symbol represents the summing function and is often drawn as a dot at the intersection of two flow lines. When the AND gate (or dot) is used, it indicates that two or more activities are performed concurrently. It also means that all of these activities must be completed before the activities following the gate can be accomplished. If the AND gate is located to the left of two or more functional blocks, it indicates that all of the functions to the right of the gate are initiated upon completion of the preceding function, as shown in Figure 2.3a. There, the two functions Drive Tool and Cool Motor must be performed when the motor is energized.

When the OR symbol is used, it means one path is mutually exclusive of the other paths, that is, you will follow only one path at any given time. To show alternative flow paths, the OR symbol must be used. It cannot be implied by unmarked gates on flow lines. Figure 2.3b illustrates this type of gate.

In some cases, there is a need to show an iterative process. This can be done with a loop as illustrated in Figure 2.4. Normal outputs are shown coming out of the right side of a functional block, whereas

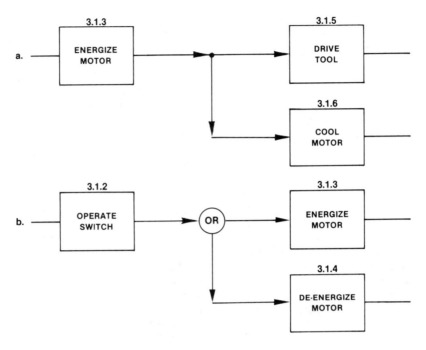

FIGURE 2.3 Flow gates. (From J. A. Burgess, Organizing Design
Problems, *Machine Design Magazine*, Nov. 27, 1969, Penton/IPC,
Cleveland, OH, pp. 120-127.)

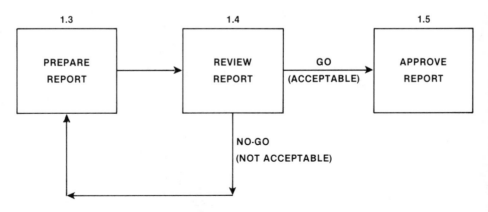

FIGURE 2.4 Interactive loop. (From J. A. Burgess, Organizing
Design Problems, *Machine Design Magazine*, Nov. 27, 1969, Penton/IPC,
Cleveland, OH, pp. 120-127.)

TABLE 2.2 Checklist for Functional Flow Diagrams

---

1. Are blocks numbered? Labelled?

2. Are interfaces shown? Can you trace the flow entering the diagram and leaving the diagram?

3. Only one input line entering the left side of a block?

4. Only one output line leaving the right side of a block?

5. Do the arrowheads clearly show the direction of flow? (Use an arrowhead at each junction and each 90° turn.)

6. Do the block titles truly specify functions?

7. Have you presupposed an answer when you labelled the blocks?

8. If so, what approaches have you ruled out?

9. Do the functional flow diagrams allow for failure situations? (Diagrams should show "what happens if . . .")

10. Are the proper gates provided for alternate (OR) and parallel (AND) functions?

11. No arrows entering or leaving top of block?

12. Are the flow lines for iterative loops or "no-go" situations shown leaving the bottom of the block? And entering the bottom of the "return to" block?

From: J. A. Burgess, Organizing Design Problems, *Machine Design Magazine*, Nov. 27, 1969, Penton/IPC, Cleveland, OH, pp. 120-127.

feedback or revision requirements come out of the bottom of the decision block and enter the bottom of the "restart" block.

From this brief explanation, it is apparent this method requires discipline as well as knowledge. Table 2.2 highlights the basic rules and identifies many of the common problems encountered when constructing flow diagrams. With a little practice, the process can be applied to many different tasks.

2.5 TRANSLATION OF REQUIREMENTS INTO DESIGNS

All of the efforts and activities described so far in this chapter are directed at identifying design requirements from many different sources. However, for these to be used effectively, the design engineer needs a way of compiling and cataloging the requirements. This section describes several techniques that have proven useful in the design process.

2.5.1 Design Checklists

One of the most common tools is the design checklist. It can take many forms, but one or more of those listed below are frequently found in use in engineering departments.

> *Department Inputs*: A simple listing with the requirements grouped by the department that specified the requirement.
> *Component List*: Requirements grouped by assembly, subassembly and major parts. This tends to bring together those requirements which are related to a specific part or subsystem.
> *Technical Specification Format*: Uses a standard or special list of topics which normally are contained in a technical specification for that type of equipment. One such listing is illustrated in Table 2.3.

The exact arrangement or format of a designer's checklist is not that important. What is important is that it is used. Developing such a list stimulates the thinking process. Even if the list is not complete as the design is started, it should be initiated and added to as the design progresses.

TABLE 2.3 Technical Requirements Topics

Basic performance	Interface requirements
Mechanical	Arrangement
Electrical	Mechanical/electrical/
Flow	hydraulic
Speed	Connection points
Etc.	
	Materials and processes
Physical characteristics	
Size	Packaging, identification and
Shape	marking
Weight	
Strength	Handling and storage
Etc.	
Applicable specifications and	Transportation
standards	
	Acceptance and testing
Operating cycle	
Life	
Startup	
Shutdown	
Emergency operation	
Etc.	

It is also a good practice to note on the checklist where each requirement originated and why it is needed. This will be very helpful in obtaining clarification at a later time or resolving conflicting requirements. It is easy to do as the requirements are being gathered initially, but it is difficult to reconstruct afterwards.

The requirements checklists are important records of the design. They should be retained in the engineering files for use in follow-on projects or for reference in this project as it progresses through manufacturing and the product goes into service. The checklist is a design basis document and is a building block not only for the design but for the later design verification efforts. More will be said about this in subsequent chapters.

2.5.2 Requirements Allocation Sheets

Another tool that is particularly useful when the functional analysis method is used is the requirements allocation sheet. This is a tabulation that follows the numbering scheme used on the functional flow diagrams. The tables are constructed as illustrated in Figure 2.5. Each block on the functional flow diagram is listed by number, and related numbers (4.0, 4.1, 4.1.1, 4.1.1.1., 4.2, 4.2.1, etc.) are grouped in succession.

The various requirements that must be satisfied by that function are then listed on the requirements allocation sheets. This is a disciplined approach and serves as a comprehensive record of the product requirements and where the requirements originated. It is particularly useful in the design of a system, since the requirements for the various pieces of equipment can be identified and allocated to the proper component. Again, these documents should be retained as part of the engineering records.

2.5.3 Incorporation into Drawings/Specifications

After the requirements from both internal and external sources have been defined and compiled, the design engineer must now see to it that each of the requirements is factored into the actual product design. This means that the drawings and specifications for the product must reflect the various requirements in words, pictures, dimensions, numbers, notes, etc.

To accomplish this, the design engineer needs to use the various checklists, customer or marketing specifications, company design standards and similar documents actively during the design process. Many of the requirements sheets will have to be given to the drafting personnel, specification writers and any other persons assigned to the project. They must use the information in the preparation of drawings, specifications, test plans and other documents which become part of the definition of the engineered product.

Component/System_____ **Sheet ____ of ____**

Block No.	Function Description	Requirements	Factors (men, money, materials, methods, machines, minutes)	Effects
4.0	Provide Supporting Structure	Structure to support power supply and prime mover.	Assembly to be light weight but sturdy in construction.	Tool must have rugged appearance
4.1	Provide Hand Grip	Must fit comfortably in man's hand and include switch for operating prime mover.	Provide insulation to prevent electric shock or burn to user.	Tool weight to be distributed for balance around hand grip.
4.2	Support Power System	Provide shock mounts to minimize vibration. Allow passages for coolant air flow over prime mover.	Shock mounts to withstand 160 F for one hour without deterioration.	N/A
4.3	Support Tool	Accommodate tool motion but keep tool within 0.005 in. of true position.	Provide sealed bearings designed for 5,000-hr life.	Minimize tool wobble to maintain accuracy. No lubrication service required.
4.3.1	Provide Tool Motion	Minimize friction in tool supporty system.	Prime mover rotation 1,750 rpm.	Minimum noise.
4.3.2	Grip Tool	Accept cutting tool with max dia of 2.000 in. or min. dia. 0.500 in. Lock tool to resist machining.	Gripping surfaces to be hardened to minimize wear.	N/A
4.3.2.1	Unlock Jaws	Provide means to free jaws when prime mover is de-energized.	May use manual locking tool.	N/A
4.3.2.2	Open Jaws	Provide means to open jaws with finger pressure only.	Remove burrs and sharp edges. Provide knurled surface for gripping.	N/A

FIGURE 2.5 Requirements allocation sheet for hand tool. (From J. A. Burgess, Organizing Design Problems, *Machine Design Magazine*, Nov. 27, 1969, Penton/IPC, Cleveland, OH, pp. 120-127.)

This simply emphasizes the need and importance of compiling the requirements in a workable form. Also recognize that the requirements sheets must satisfy two different purposes. The first is to be the input data to the design process and do it conveniently and efficiently. The second purpose is to be the yardstick for evaluating the design to determine if it meets all of the applicable requirements. Even though persons other than the design engineer may contribute significantly to the design of the product, it is usually the responsibility of the design engineer to see that all of the requirements were incorporated. The checklists are especially useful in this verification effort.

As will be described further in later chapters, the design requirements are a fundamental element in any design assurance program. Time spent in understanding the real requirements for the design and compiling them in an effective manner can contribute significantly to the success of the final product.

2.6 SUMMARY

Design requirements are the building blocks of product design. Yet, some organizations seem to slight the process of defining the require-ments in detail, understanding the impact each requirement has on the design, and collecting and documenting the requirements in a way that the design team (or single designer) can use easily.

It must be recognized that each product is eventually judged by its users against their own set of requirements, however complete or foggy. Thus, it is the challenge of the design group to anticipate and understand the product requirements, factor these into the designs and monitor compliance with the requirements as the product moves through the various stages of development, production and operational use. And it needs to be done early and in a disiplined manner to have the best possible effects on the product design.

3

Drawing Control

3.1 Introduction   33

3.2 Types of Drawings   33

    3.2.1 Layout Drawings   33
    3.2.2 Detail Drawings   36
    3.2.3 Assembly Drawings   38
    3.2.4 General Arrangement Drawings   41
    3.2.5 Special Drawing Types   42

3.3 Drawing Preparation   46

    3.3.1 Drawing Cycle   47
    3.3.2 Drawing Requests   48

3.4 Drawing Review and Approval   50

    3.4.1 Draftsman Review   51
    3.4.2 Engineer Review   51
    3.4.3 Technical Specialist Review   51
    3.4.4 Technical Team Review   52
    3.4.5 Choosing the Right Method   52

3.5 Drawing Document Control   53

    3.5.1 Drawing Numbering Systems   53
    3.5.2 Drawing Release System   55
    3.5.3 Drawing Distribution   56

3.6 Drawing Revisions   57

    3.6.1 Change Requests   57
    3.6.2 Revisions to Drawings   60

3.7 Summary   62

3.1 INTRODUCTION

Drawings are the fundamental tool of engineering departments. In
fact, drawings are typically the engineer's most important product.
Stated in terms of production, drawings comprise the major portion of
the output from an engineering organization.

Since drawings are so common to the engineering process, it is
easy to conclude that all engineers know how to prepare and properly
control drawings. However, experience shows this just is not true.
Many problems occur in industry because of incomplete, inaccurate,
or unclear drawings.

The worst troubles frequently occur because the engineers and
drafting personnel are very familiar with the design and their intent
but fail to communicate the design requirements to the users of the
drawings. Missing or ambiguous notes, unspecified tolerances, lack
of dimensions, unclear views and incomplete drawing changes are common examples of problems encountered daily in many organizations.
Some of the problems show up early and simply introduce delays in
production. Other problems may be more subtle and are not found
until the parts don't fit or the unit fails to operate properly. The
latter problems are, of course, more serious and may be costly to
correct.

Thus, the development of an effective drawing control system is a
starting point for nearly all design assurance programs. This chapter
describes several types of drawings commonly used to define a product.
Information on the basics for drawing preparation, review and release
is presented in terms which stress those factors that enhance the quality of the drawing effort. Also, recommendations for controlling the
distribution of drawings and the control of drawing revisions are given.

3.2 TYPES OF DRAWINGS

Drawings are prepared for one primary purpose: to convey information. As might be expected, this can be accomplished in many different ways. Over the years, the number of different types of drawings
has grown with the introduction of new technology. New and unusual
types of products have created a need for new and unusual types of
drawings. Nevertheless, there are a few basic types which apply to
most products and industries. This chapter focuses on these basic
types and provides information regarding preparation and control.
The same principles can be applied to other specialized types of drawings, as needed for a particular product or application.

3.2.1 Layout Drawings

The term, layout drawing, is used to describe an overall design drawing. It is used in the early phases of the design process to record

the size, shape and arrangement of the product. It is part of the
creative process in engineering design. The layout of the product
may be full size, expanded size or reduced-scale size, depending upon
the needs for clarity and comprehension.

Engineers normally start with some simple, and often crude,
sketches of the new or modified product. The sketches and some further descriptive information are then used by the layout draftsman to
create the design in proper proportion and scale. Figure 3.1 is an
example of a layout drawing.

The layout drawing has several characteristics which makes it different from other types of product drawings. These differences are
significant and need to be recognized for proper design control. The
key characteristics of a layout drawing are:

The pictures are drawn precisely to scale.
Only important or major dimensions are shown.
Tolerances are seldom listed—only on key or critical dimensions.
Cutaway or unconventional views are frequently shown on layouts.
Bills of material are generally not used, or are used only to a
    limited degree.
Materials and process specifications are often not listed.
Notes on layouts generally explain a design feature and are not
    for production use.
Layout drawings are frequently changed during the design proc-
    ess, and individual changes are rarely documented.
The layout drawing is often allowed to become obsolete or super-
    seded as the design moves into the production phases.

These differences present some special opportunities and problems
from the standpoint of design assurance.

The layout drawing can be a very useful tool in exploring many of
the physical size, shape and arrangement aspects of a new design.
It's quite inexpensive to make changes on paper. However, the ease
of changing a layout drawing and the typical lack of documentation of
the changes make it difficult to use it as a method of control. Layout
drawings tend to stay on the draftsman's board throughout much of
the design phase. Working prints are occasionally produced, but the
layout generally is not officially released, or, at least, not until
virtually all design work has been completed. Nevertheless, accurate
layout drawings can be very useful for checking not only the design
features but also serving as the baseline for checking the production
drawings as they are being prepared.

One of the techniques for giving visibility to the status of a layout is to add a design control table near or adjacent to the layout
drawing title block. Each time the layout is removed from the board
to make copies, the draftsman enters a release number or revision
letter and the date (and time) of the release. An example is shown in

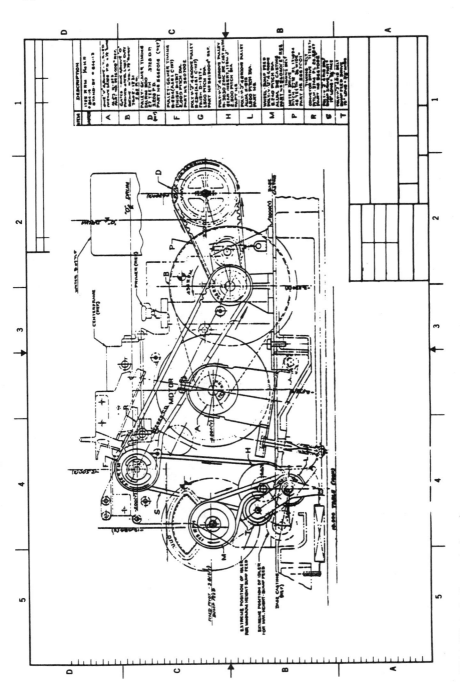

FIGURE 3.1 Layout drawing.

Figure 3.2. This process is not foolproof, but it helps the engineering personnel keep track of the drawing status as they work with the prints.

3.2.2 Detail Drawings

A detail drawing is used to define an individual part or piece for manufacture. It contains all of the specific requirements for size, shape, materials of construction, joining, finishing, etc., to produce the item, as shown in Figure 3.3. There are many standard references available which describe the proper methods and techniques for constructing detail drawings for production and that information will not be duplicated here.

Various aspects of detail drawings need to be addressed as part of the design assurance program. Consider these factors.

All dimensions must be toleranced. Show the tolerances with the dimensions or specify standard tolerances on the drawing, e.g., such as illustrated in Table 3.1.

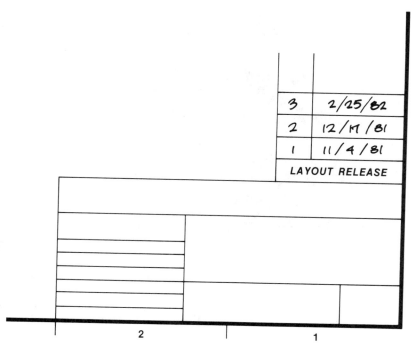

FIGURE 3.2 Layout release control block.

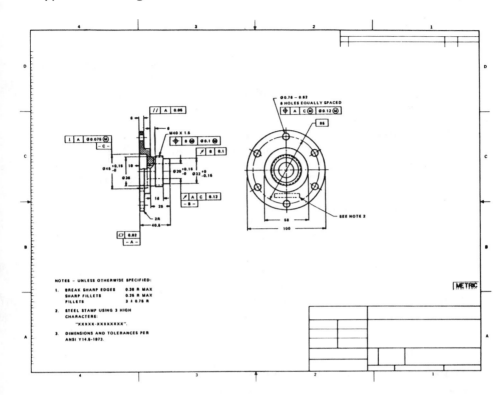

FIGURE 3.3 Detail drawing.

TABLE 3.1 Standard Drawing Tolerances
(unless otherwise specified on the drawing)

Type of measurement	Tolerance (inches)
Linear—one decimal place	±0.06
Linear—two decimal places	±0.020
Linear—three decimal places	±0.005
Angles	±1/2 degree
Inside radii	0.015 min
Outside radii	0.015—0.030

Define the materials of construction to be used in clear and specific terms. Wherever possible, use industry-recognized specifications or company specifications. Otherwise define the materials by generic name and critical physical/chemical properties,
e.g., cold-rolled steel, 0.03 carbon max.; phosphor-bronze,
100,000 psi minimum tensile strength.

Define required special processes, e.g., welding, plating, paint-
ing, etc., in clear and specific terms. Wherever possible, use
industry-recognized specifications and standards. Where it is
necessary to define special or critical requirements, specify
definitive parameters and acceptance criteria. For example,
"stress relieve at 900-1000°F for 2 hours;" "liquid penetrant
inspect surface B after final machining. No cracks allowed per
visual inspection."

Review all instruction notes carefully. Are they clear and free of
ambiguity? Are acceptance criteria specified where needed?

In general, these four areas, tolerances, materials, processes and
instructional notes, cause the vast majority of problems on detail drawings. Proper attention to these can bring big returns in achieving the
design performance and minimizing the chances for errors in manufacturing and use.

3.2.3 Assembly Drawings

There are many types of assembly drawings, but all serve the same
purpose. They show what parts go together, how they are joined and
how many of each are required. Simple as it sounds, many assembly
drawings fail in their mission. The drawings just do not convey the
necessary information to the persons that must put the pieces together
to construct the desired assembly.

Figure 3.4 shows a common type of assembly drawing. It contains
many of the basic elements found in assembly drawings; e.g., bolted
assembly, finishing, bill of materials, instruction notes, and acceptance
criteria. It is the combination of requirements that seems to cause so
much trouble.

Assembly drawings must convey a lot of information and do it accurately. The drawing arrangement must reasonably resemble the
actual hardware, but it does not have to be as precise or accurate as
does the layout drawing. Nevertheless, care should be taken to draw
the various views in approximate proportion to the real pieces. Otherwise, the apparent difference between picture and part leads to confusion, "creative" assembly and outright errors in the final hardware.
Obviously, all such problems and errors must really be attributed to
shortcomings in the design process.

One of the frequent causes of this series of events is insufficient
attention to detail. Design engineers often feel the assembly drawing

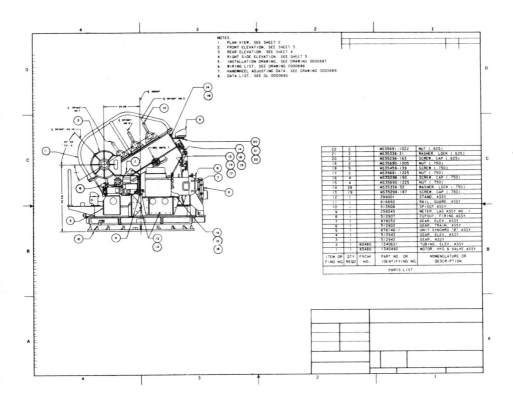

FIGURE 3.4 Assembly drawing.

is not worthy of their attention, and it is left to or delegated to (sometimes forced upon) a draftsman or technician, who may not be familiar with shop or field practices and problems. However, some engineering departments have been particularly successful in overcoming problems with assembly drawings. They accomplish this through a thorough investigation of their problems and then documenting the preferred practices. Many of the proven practices are summarized below.

Identify each component clearly in the bill of material by part number and/or drawing item number.

The total number of parts required must be specified, and the drawing views must show where all of the parts listed in the bill of material are physically installed. There are a few exceptions to this rule, e.g., where some items must be used as required to fit the particular unit, such as lockwire, weld metal, etc. However, it is important to minimize the use of "As Required" on assembly drawings to improve effectiveness in the assembly and operational phases.

Strive for clarity in instructional notes. Frequently the notes are among the most important features on the drawing. Although there is a tendency to write drawing notes with the minimum number of words, like a telegram, drawing notes must convey information correctly to the user. Unfortunately, brevity often leads to lack of clarity.

Relate instructional notes to a particular operation, sequence or part. This can be achieved by specifying the affected drawing items in the text of the note, such as, "Torque bolts, item 22, to 25-35 foot-pounds."

Specify the methods for joining parts with care. There are many different ways of joining piece parts to form the next-higher-level of assembly. It may be by welding, brazing, soldering, bolting, clamping, etc. Each method has certain special characteristics that must be defined on the drawing or by reference to some other specification or instruction.

When welding, brazing, soldering, or similar joining process is involved, use the correct symbols, and define the applicable process by recognized specification or standard designation. Be sure the location, type and size of joints are indicated on the drawing. Incorrect or incomplete weld symbols are frequent causes of problems.

Specify applicable acceptance criteria for the various assembly processes. The following are representative examples of:

Dimensional: 0.001-0.003 radial clearance between items 7 and 6.
Physical: Material hardness to be Rockwell C 24-32.
Operational: Switch must close in 0.5 seconds max.

The criteria provides the basis for determining an acceptable unit from a non-conforming one.

Provide all of the necessary detail views to show where and how different parts fit and must be assembled. This is particularly troublesome when there are right-hand and left-hand or mirror-image combinations. Special views are often needed to show different fit-up conditions, use of shims or spacers to obtain needed clearances, etc. For some reason, many draftsmen like to use modification notes or drawings rather than preparing an extra view or additional subassembly. Modification notes or drawings tend to say such things as: "Make it like item 1 but don't drill the hole in it." Although this process may reduce the drawing time significantly, it increases the error potential in a similar proportion.

In summary, assembly drawings are a major source of errors. Engineering departments tend to underestimate the importance of preparing correct and complete assembly drawings. Any organization wanting to

enhance the quality of their design effort should examine assembly
drawing practices thoroughly.

3.2.4 General Arrangement Drawings

General arrangement drawings frequently look like a simplified layout
or assembly drawing. However, the purpose of a general arrangement
drawing is different from these other types. It is intended to show
potential users what the exterior of the finished product looks like in
terms of its external shape and size and provide key information about
mounting points, connections, and perhaps, gross weight, center of
gravity, or other characteristics important to the user. Figure 3.5
illustrates a typical general arrangement drawing. This type of drawing is also referred to as an outline or envelope drawing.

The general principal to be remembered when preparing or revising a general arrangement drawing is: What does the user need to

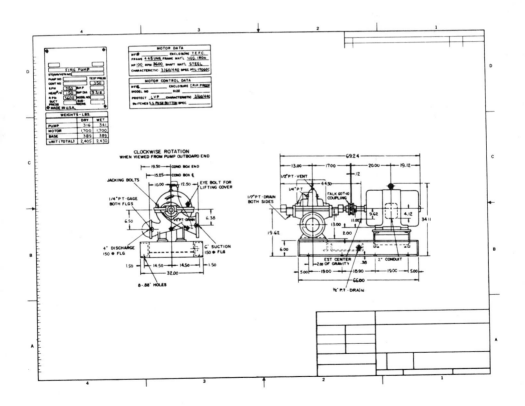

FIGURE 3.5 General arrangement drawing.

know from this drawing that will increase the probability that the component will be handled, installed and operated correctly for its intended use?

The picture portion of a general arrangement drawing needs to be a reasonably accurate illustration of the actual hardware. Nothing is much more confusing than having a picture that looks noticeably different from the actual component.

Overall dimensions, e.g., length, width, height, should be shown with tolerance or maximum/minimum values, as appropriate. The type and size of connections, e.g., mechanical, electrical, hydraulic, etc., must be specified in terms that can be used to select and attach the mating parts. Attachment points for mounting, such as, pads, flanges, or brackets, must define the details of hole size and location, thread type and size, bolt circle diameters or similar mounting information. Keep in mind the impact on the users of not having the necessary information to use your design correctly in their application.

3.2.5 Special Drawing Types

Although there are many, many different and specialized types of drawings, there are a few others that are particularly useful to the design engineer as part of the design assurance effort. Each of these types will be briefly described below. The special contributions of, or concerns about, these drawings to the design assurance effort are also explained.

Specification Control Drawings

Variations of this type of drawing are also called a procurement control or source control drawing. It consists of one or more external views (similar to a general arrangement drawing), a few important dimensions or physical characteristics and a table which contains selected manufacturers' names, addresses and part numbers. An example is shown in Figure 3.6. A specification control drawing is used when only one or two vendor's products are known, although other equivalent products may be acceptable.

In contrast, a source control drawing is used when only certain products have been proven acceptable for this application and are the only ones permissible to use. The benefit of this approach is to restrict procurement to only those sources listed on the drawing. The drawing normally carries the notation that only the vendor part numbers listed are acceptable (no "equivalent" parts may be substituted).

The engineer should recognize that the use of a specification control drawing or a source control drawing does not prevent the vendor from making changes to the part, but it frequently does reduce the chances of problems caused by the surprise introduction of a new and possibly unqualified source of supply. On the other hand, limiting

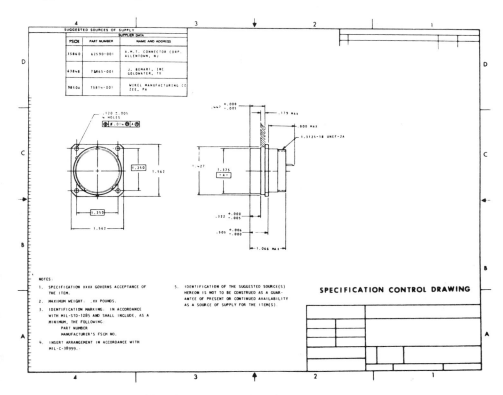

FIGURE 3.6 Specification control drawing.

the sources of supply may reduce cost competition and limit the purchasing department's ability to negotiate cost improvements with suppliers.

Specification control and source control drawings have a useful place in critical applications or for new development projects. However, this process should not be used to an excess.

Book Form Drawings

A book form drawing is best described as a word drawing. It is usually prepared on standard drawing formats and may consist of several sheets. Book form drawings frequently are used to convey assembly or installation instructions. The first sheet of the book form drawing serves as title page, and revision column. Subsequent sheets containing the index, sketches and narrative text to define the technical requirements, as illustrated in Figure 3.7.

Drawing rooms typically dislike book form drawings, because they are a major departure from conventional drawings. Also the multi-sheet

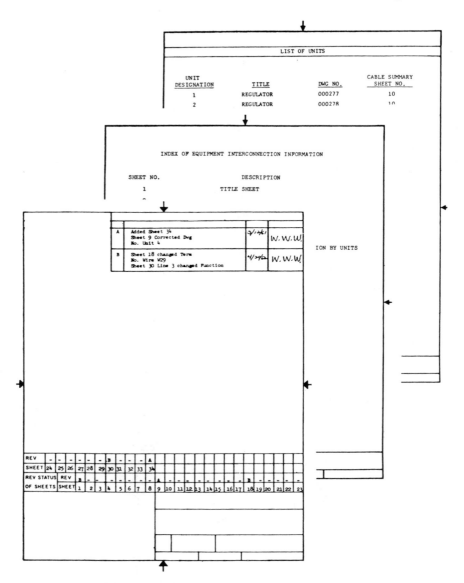

FIGURE 3.7 Book-form drawing.

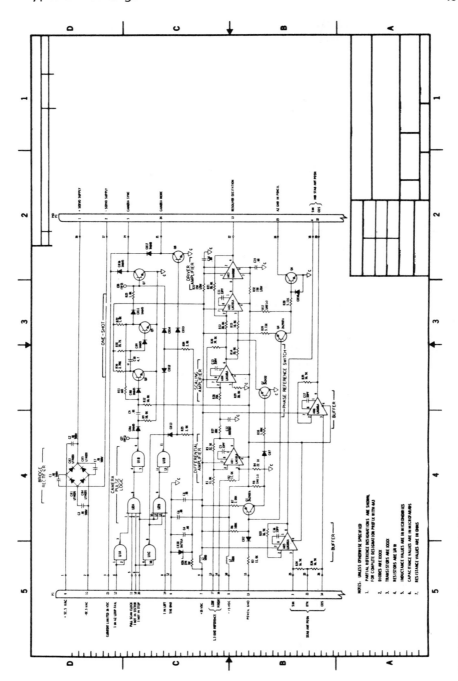

FIGURE 3.8 Electrical diagram.

format is often cumbersome for conventional drawing reproduction
facilities, and the text of the drawing is usually prepared by type-
writer which further complicates the revision process.

Nevertheless, selective use of book form drawings is appropriate
to consider. This type of drawing may offer benefits when an engineering department faces a rigid or externally controlled specification
system. It may allow a needed control or flexibility that cannot readily
be obtained by using conventional memoranda or specifications. Book
form drawings can be identified by standard drawing numbers or special numbers can be assigned, according to local preference. It is a
document system that can be completely controlled by the engineering
organization.

Electrical Diagrams

Although wiring diagrams are a very common form of drawing, there
are some special features worth discussing in the context of design
assurance. All electrical/electronic equipment are described in some
manner by a wiring diagram, schematic or block diagram. The most
significant characteristic of these types of drawings is the functional
representation presented. Figure 3.8 illustrates this feature. Electrical drawings typically are arranged for ease of presentation and
do not necessarily look like the actual hardware configuration. The
use of blocks and single lines show the functional arrangement, but
that is often different from the physical configuration, e.g., several
different connections may be contained in one multi-strand cable.

The sheer number of functional connections of a typical electrical
schematic or wiring diagram make these drawings a frequent source of
errors. It is easy to miss a junction, fail to notice an unconnected
part of the circuit, or incorrectly identify a component. The result is
often improper operation or failure to operate.

These drawings require careful checking and close attention to
detail. The simplicity of single-line representations can easily become
the complexity, or error, factor. Don't treat electrical drawings lightly, or many problems can occur rapidly.

3.3 DRAWING PREPARATION

Except for the very smallest of manufacturing organizations where the
designer makes his own drawings, persons other than the design engineer normally prepare the various types of drawings. This requires a
means for defining not only the work to be performed but also the administrative aspects of controlling the cost and scheduling of it.

3.3.1 Drawing Cycle

The usual sequence of a design project is to start with a concept, probably defined with some rough sketches. A draftsman then converts the originator's ideas into layout or schematic drawings. These provide the engineering working drawings for detailed investigation and analysis of the design. The design drawings often go through numerous changes as the ideas for a new or modified product evolve. These drawings may even be used for the construction of models or pre-production parts for testing and analysis.

After the decision is made to proceed with production of the new design, detail drawings of the numerous parts and pieces must be prepared. These drawings will then be used by the factory and outside suppliers in the manufacture of the individual items which make up the end product.

As detail drawings are prepared and completed, various kinds and levels of assembly drawings must be prepared to show how the various individual parts must be combined and joined to construct the completed product. This cycle is shown in Figure 3.9.

Methods for accomplishing these tasks vary with the size of the engineering organization, the complexity of the product, and the nature of the business. Nevertheless, several basic methods are

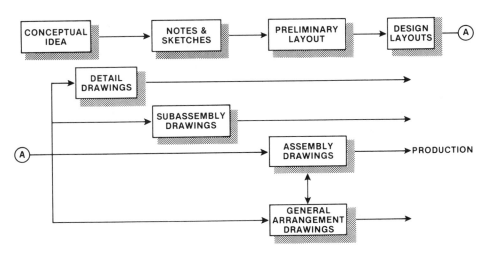

FIGURE 3.9 Typical drawing preparation cycle.

common to nearly every type and size of engineering organization.
These are described in the next section.

3.3.2 Drawing Requests

A drawing request is used, as the name implies, to request the preparation of a new or revised drawing. In some firms, a drawing request
describes the work to be done in considerable detail. In other companies it may be little more than the vehicle in the paperwork system to
identify where to charge the labor hours for cost accounting.

For design assurance purposes, the drawing request needs to be
used in a disciplined manner to define the scope and purpose for the
drawings, the technical requirements to be incorporated or satisfied,
and the special features to be included, in addition to the ever-present
administrative details for scheduling, project identification and charging
time.

One large firm has identified its drawing request process as a
design specification. The design specification, commonly called "D
Spec," serves as a written record of the design project. A sample
design specification is shown in Figure 3.10. It is prepared initially
by the design engineer assigned to the project and is the kickoff document to start the drafting work. The Drafting Department creates a
new job file and assigns the D-Spec a unique identifying number.
Figure 3.11 illustrates a typical cycle for controlling D-specs.

For design assurance purposes, D-Specs should be reviewed and
approved by a person, other than the originating design engineer, who
is technically knowledgeable and competent in the field. The purpose
of this review is to provide a technical check-and-balance process
early in the design cycle. It is often performed by the engineer's
supervisor, or lead engineer, if it is a team effort.

As the design work progresses, the design engineer supplements
his instructions to the Drafting Department by issuing additional
D-Spec sheets. Sketches, technical requirements, marked up drawings, etc., are attached to the D-Spec. The supplement also receives
a technical review and is then given to the assigned draftsman. The
initial D-Spec and any later supplement sheets are retained in the
D-Spec file folder for the project. This provides a useful record of
the design process and tends to improve the quality and the completeness of the engineer's instructions to the Drafting Department.

Written design requests do not prevent the day-to-day informal
conversations and verbal instructions between the engineer and draftsman, but the D-Spec system does add an element of discipline to the
technical communications process. The use of written instructions
during the design phase avoids many problems caused by misunderstandings of verbal instructions.

SUBJECT		
Layout for Air Handling System for Bldg 204		

CUSTOMER		JOB NO.
W.W. Wolfgang Enterprises		8015

DESCRIPTION

      Prepare layout for a ceiling-mounted air handling system. Size for both summer air conditioning and winter heating volumes. See Engineering Sketch 83-0720JB attached for duct routing.

      Air handling units to be roof mounted. Plan for 4 air changes per hour. Use replaceable fiberglass-type filters. Gas-fired heating units to be located inside near ceiling. Air conditioning to be integrated with air handling units. Use computer-controlled, air-actuated dampers for balancing and flow adjustment.

CHARGE NO.	ORIGINATOR J. Nicole
8015 - 271	APPROVED C. Lynne
REQUESTED COMPLETION DATE	
8-31-83	APPROVED R. Nichols
D-SPEC No. 47-302	DATE RECVD 7/21/83

FIGURE 3.10 Design specification form.

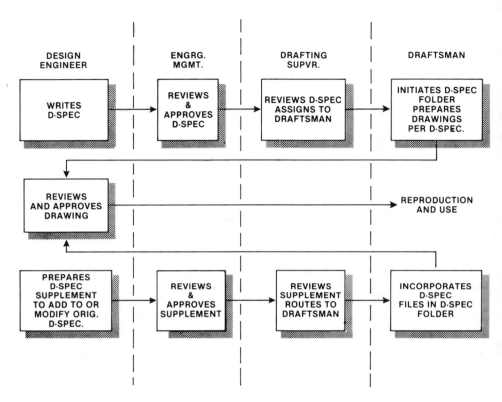

FIGURE 3.11 Design specification cycle.

3.4 DRAWING REVIEW AND APPROVAL

There are several schools of thought, regarding what is the best way
of verifying the adequacy of drawings prior to releasing them for hardware use. Many different approaches are found in industry, and several of these are described in the section. There really does not seem
to be "one best way," but there are some important fundamentals that
need to be addressed, regardless of the approach selected. Again
the complexity of the product and the criticality (degree of risk and
consequences of failure) of the application should be considered when
choosing the review and approval process. These influences may dictate the need to have persons with specialty knowledge review and/or
approve the drawings.

    Several categories of information that should be examined in the
review and approval of product drawings are listed below:

Compliance with technical requirements, including geometry.
Selection and definition of materials of construction to be used.
Selection and definition of special manufacturing processes.
Ability of the manufacturer to make and check the part or assembly.
Definition of tolerances and acceptance criteria.

There are several different ways that a drawing review can be accomplished. Some of the more common approaches are described below.

3.4.1 Draftsman Review

There are many firms that use the draftsman as the sole drawing preparer and reviewer. The basis for this is that the draftsman is the one most familiar with the drawing methods and techniques and is in the best position to convert the engineer's information into proper and workable drawings. Also many firms consider it inappropriate for their high-paid engineers to be worrying about the details that a lower-classification person can handle. The possible shortcomings of the draftsman review is the potential of overlooking one's own mistakes or omissions or limited knowledge of factory or operational conditions. One variation in this approach is to have the drawing reviewed by a second draftsman, possibly a full-time checker or layout draftsman. This provides a second set of experienced eyes to check accuracy and completeness.

3.4.2 Engineer Review

Another approach used by a large number of firms, particularly those with small-to-medium sized engineering departments, is to make the design engineer responsible for his own drawings. The drawings are prepared by drafting personnel, but the originating engineer must review and sign off the drawings. This process provides second party review (draftsman and engineer) by persons who are very familiar with the design and application requirements. However, this same familiarity sometimes leads to oversights or over-simplification caused by lack of objectivity in the review. Nevertheless, the engineer review approach works reasonably well.

3.4.3 Technical Specialist Review

To overcome a lack of specialty knowledge or enhance the skills of a relatively inexperienced engineer, some firms introduce a second technical person to review and approve the design engineer's drawings. This person may be the engineer's supervisor, design team leader, or a senior engineer who is knowledgeable and experienced in the products covered by this design. The technical specialist serves as an objective reviewer and looks for errors, omissions and deviations from proven practices. This process obviously adds something to the time and cost of drawing preparation. But those who use it feel the approach pays for itself in the long run by avoiding later problems in production and operation.

3.4.4 Technical Team Review

Large firms, especially those working in high-technology areas, commonly use a team of technical specialists to review and approve the drawings for each new project. This approach is based on the premise that no one person is adequately qualified to be knowledgeable of the many different technologies involved in the product to provide an adequate review of the drawings. Various specialists are then delegated the responsibility to review and approve the product drawings. Typical specialty disciplines including in the drawing review/signoff process include materials engineering, manufacturing engineering, quality/reliability engineering, and structural, thermal or systems engineering, in addition to the design engineer. Use of the review team is common on military and aerospace projects.

3.4.5 Choosing the Right Method

Obviously, there is no one best way for reviewing and approving drawings for all situations. These must be tailored to the local needs of each engineering department and type of product. Nevertheless, experience shows that the quality of design can be enhanced by following certain practices. For example, where a new product involves new technology or significant changes from local manufacturing experience, it is a recommended practice to bring the best skills available into the drawing review process. This particularly applies to the review and approval of new concepts as shown on layout drawings. Getting other departments involved early in the design process also stimulates thinking and planning for supporting follow-on activities in those other departments. In addition, many firms routinely require product layout drawings to be signed off by the engineering manager or the project manager before the design can be released for production. This is to ensure technical management involvement and support.

Companies that are in a job-shop type of business tend to rely heavily on the design engineer as the authority for approving the design drawings. This is especially true where many details are involved for a custom-tailored design, but the basic design tends to be reasonably fixed, such as in the case of generation and distribution equipment for the electric utilities. In this case, the engineer may draw upon the expertise of available specialists and consult with them as deemed necessary. However, the design engineer makes the final decision of what appears on the detail drawings.

For assembly drawings, particularly those for.the major assemblies and for customer use, it is a good practice to have more than just the design engineer involved in the drawing review and approval. Again, this approach is most applicable to products which are complex or are used in critical applications. Review and signoff by one higher level of engineering management and one or two specialists from manufacturing and materials/processes groups will typically provide the appropriate specialty knowledge to assure the desired level of quality.

Experience shows that multi-disciplined drawing review works well and does not need to add significantly to either the time schedule or cost of a new product. It is particularly cost-effective when the reviewers conduct an on-going, informal review in the drawing room as the drawings are being prepared. Then at the time for the final review and signoff, there are seldom any surprises or delays.

3.5 DRAWING DOCUMENT CONTROL

There are three major elements to the document control process for nearly all types of engineering documents, and these are particularly visible for engineering drawings. The key ingredients are: (1) how to identify individual documents, (2) how to authorize their use and (3) how to get the proper documents in the hands of the users. Each of these aspects of document control will be discussed in subsequent sections.

3.5.1 Drawing Numbering Systems

The purpose of assigning drawing numbers is to give each drawing its own unique identification. Many companies use or encounter hundreds, even thousands, of different drawings in the course of conducting their work. Thus, some means must be available for telling one drawing from another.

There are many different types of drawing numbering schemes used in industry. However, the vast majority of these can be grouped into three general categories: Non-significant, partially significant and significant numbering systems.

The first category, non-significant numbers, is simply a type of serial number which is assigned in the order the drawings are created. For instance, the first drawing is assigned 10,000, the next 10,001, then 10,002 and so forth. Number assignments are made independent of drawing size, type or application. The exact number of digits or combination of letters and digits is entirely up to the judgement, preference and needs of the organization. It should be recognized that, although a non-significant drawing number is easy to administer, it is also prone to human error, such as transposition, duplication, etc. Therefore, care must be exercised throughout the total organization in the accurate use of the drawing numbers.

Another type of drawing numbering scheme is tied to drawing size. This is a partially-significant numbering system. The drawing number typically includes a letter which corresponds to the common, standard drawing format sizes; A for 8-1/2 × 11 inches, B for 11 × 17, C for 17 × 22, D for 22 × 34, E for 34 × 44, J or R for 44 inches wide and with lengths up to 8-12 feet or more (also referred to as roll-size drawings). A drawing number in this system might look like: 23C4673 or 542B916 or E918485. In each case, the letter represents the format size of the drawing. This system is primarily designed to enhance filing of the different-sized original drawings in the drawing vault. It may have some benefit for recognition in that certain types of parts tend to be drawn on one or two drawing format sizes. However, this does not hold universally true. Another type of partially significant drawing number is to use a special prefix to designate that certain drawings apply to a particular product project. However, this tends to restrict the broader use of the drawings.

The third type of scheme is a significant numbering system. In a significant numbering system, the drawing number itself conveys information as well as unique identification. This information might be the product type, model number, originating plant or applicable project. Engineers tend to favor significant numbering systems, because of their inherent logic and form of technical shorthand.

If a company produces numerous variations of the same basic product design, it may be possible to assign the same prefix or suffix number for the same type of component or data. For example, drawing number XXXX-01 is always the outline drawing, XXXX-02 always the top assembly drawing, XXXX-04 the wiring diagram, etc. The project or model number is then used as the base number. Two different examples of this method are shown below.

EAM	1664	01
Product Code	Order No.	Drawing Type
		(Also serves as the unique drawing number for the application)
7844	672	48
Project No.	System/ Subsystem Category	Drawing No.
		(Not necessarily related to a particular type of drawing)

Although a significant numbering system can be very helpful in communications, it does not fit all situations. The most obvious problem is the difficulty in assigning numbers to newly created designs, or using them in applications where the product configuration or project content varies considerably from job to job.

In all drawing numbering systems, the integrity of the system depends heavily on the controlled assignment of the numbers to prevent duplication. Skipping a number or group of numbers is not a problem, but two different drawings with the same number is. From the standpoint of design assurance, steps need to be taken to minimize the chances of duplicating drawing numbers.

One approach is to use a single drawing number list which is controlled by the drafting department. As a drawing is initiated, the assigned draftsman obtains the next applicable number, and the necessary identification information (date assigned, drawing title, draftsman's name, etc.) is recorded on the drawing number list. Another variation of this is the use of pre-numbered cards. When the draftsman starts a new drawing, the administrator of the drawing number system requires the draftsman to fill out the drawing number card. The card remains in the possession of the system administrator while the draftsman prepares the drawing. Then when the drawing is completed and ready for use, the administrator submits the card and the original drawing for reproduction and vault storage. The drawing number card is then used to update the list of completed drawings.

With the spread of computer-aided-design and the extensive use of computers in engineering departments, more and more engineering organizations use their computers to control their drawing number assignments. This will be discussed further in Chapter 5 as part of the overall approach for controlling product identification in the factory and in operation.

3.5.2 Drawing Release System

Another important part of the document control process for drawings is the method for authorizing use of the drawing. Some firms are small and simply put a print in a job folder and give it to the factory group to use. Other companies give the drawing to the reproduction services to make copies and to route to predesignated areas or departments. Upon arrival, the drawing is then available for use. Other companies, especially those making complex products, or those having many manufacturing sections, tend to use a drawing release form. This is a written announcement to specified individuals or departments that a particular drawing or group of drawings is now available and authorized for use. The document may have many different names, such as, Engineering Drawing Authorization, Drawing Release or Engineering Order, but its purpose is the same: to announce the availability of a drawing and to give authorization for its specific application.

From a design assurance standpoint, the drawing release process is an early attempt to to control the design. It tells others what drawing to use for what purpose. Some companies use the drawing release process for advance ordering of long-lead-time items. This is generally a valid and workable arrangement if limited to a small number of items. Others use it to permit procurement of all detail parts to begin before completed assembly drawings are available. However, the latter practice is risky, since this approach makes it easy to overlook some parts or result in the ordering of incorrect quantities.

Wherever possible, it is a preferred practice to release a complete assembly at one time. This includes the various assembly, subassembly and detail drawings required to make and join everything up to that level of assembly. Such a practice minimizes the chances of omissions or quantity errors.

3.5.3 Drawing Distribution

The important step after engineering finishes the necessary drawings is to distribute the drawings to those groups that need them. Although drawing distribution may seem like a minor detail and not related to the engineering process, it can contribute to, or detract from, the design intent. Many organizations have discovered that a piece of equipment did not work properly because an incorrect drawing was used. It is not uncommon to discover that the factory or the field is working to a superseded version of a drawing.

To avoid these problems, each firm must define who needs what types of drawings and then arrange to distribute the drawings and all subsequent changes in a timely manner to these groups. The distribution system needs a few specific provisions to be effective. First of all, there needs to be one central point for control. Frequently, this is a reproduction service facility. The control point must maintain an official list of drawings under control, what revision is applicable, and where copies of the drawing have been sent. In large facilities there may have to be satellite facilities that maintain the records of localized distribution. However, if there is ever a question of what drawing number and revision is to be used, the central control point must be the final authority.

The control point must know how many copies of the various drawings need to be distributed to which locations. Then as revised drawings are issued, the control point must be capable of getting the latest information to the designated groups promptly. This can be accomplished by simply mailing a copy of the new revision to all groups designated to receive the original release. This system works reasonably well in small to medium-sized firms. Another approach is to issue a standard notice (or a copy of the drawing release form) to the specified groups. Then those persons are responsible to order or pick up the latest drawing. This is often used in very large organizations, but it does depend upon the persons receiving the notice to take their own actions. As a result, it may be only marginally effective.

A more positive means which contributes to design assurance effectiveness is to have the control point physically distribute the new drawing or revision to the designated locations, replace the superseded drawing with the new drawing, and destroy the superseded version. Although this obviously requires more resources than the other two approaches, it is often justified in plants which build complex products or products for applications which must be closely controlled.

Some firms distribute two types of prints of drawings. The first type is stamped "Controlled," and all such prints are used to design and build the product. These go through the formal drawing distribution system as described above.

The second type of print is stamped "Uncontrolled." It is issued upon request and is the latest revision available at the time the print was produced. It is intended to provide information on a "now" basis. No effort is made to supersede the uncontrolled print when a later revision is released. Thus, persons using uncontrolled prints should use it for their immediate need and then discard it. If this approach is used, people in the organization must understand the difference and significance of using controlled and uncontrolled prints and use them correctly in a disciplined manner.

The basic principle in all of this is the drawing distribution system must provide an effective means of getting the current drawings to those persons who must work with the information in a timely manner. Anything less than this can hurt the integrity of the design.

3.6 DRAWING REVISIONS

The previous sections described the processes for preparing, approving and distributing new drawings. However, a need for changing the drawings inevitably occurs. All such revisions need to be controlled as part of the design assurance effort.

Various methods for managing the control of changes are covered in Chapter 5, but some of the details relating to the revision of drawings are described below.

3.6.1 Change Requests

Some form of written documentation should be used to describe what changes are needed. This is especially important, since there are often many details that must be addressed correctly. It is not the place to depend upon verbal instructions.

The forms used to initiate a change to a drawing may be called by various titles, such as Engineering Order, Revision Notice, Drawing Change Request, Design Specification Change, etc. Figure 3.12 illustrates one example of a representative document.

Several types of information are needed on the change request to increase the probability that the originator's intentions will be properly carried out. The first is the basic identification information. The particular drawing, or drawings, to be modified must be defined by the correct title and number.

The purpose and reason for the change should also be briefly described. This will be a help to the drafting personnel to understand the intent of the change and to recognize possible subtle oversights or omissions that should be investigated. The originator of the change may not have recognized all items that must also be addressed to make the change complete. The reason and purpose are also important to help other departments that may be involved in the evaluation of the change request to understand what is being done and why. And, finally, the statement of purpose and reason for the change is very useful at a later time when there is a need to investigate a factory or operational problem, and the drawing revisions are being reviewed as a possible source of clues into the cause of the problem. It is surprising how quickly engineers and draftsmen loose track of the purpose and reason for a particular change, especially in projects where there is a high level of change activity.

The actual change must be described in sufficient detail that the drafting personnel can understand what must be done to accomplish the change. Depending on the type of drawing to be revised, and the stage of the engineering effort (development, initial production, mature product, etc.), the amount of detail in the description of the change may vary. For example, a change to a layout drawing will frequently introduce a new approach or concept and may require a considerable amount of information. On the other hand, a change to a particular detail drawing may be very specific and require only the briefest of descriptions to be understandable. Regardless of the complexity or magnitude of the change, avoid verbal instructions. Describe the change in writing to get the benefit of clarity of instruction and traceability of actions for design assurance purposes.

A change request will also serve an administrative purpose, such as identifying the budget or project to charge the drafting labor hours. This may be important for cost accounting, but do not use a change request document solely for that purpose. Let it provide a brief history for future reference of what was done and why.

It is also a good idea for design control purposes to require each drawing change to be reviewed and approved by another knowledgeable technical person prior to releasing it. This is particularly appropriate in the early stages of design and can be accomplished by the

DWG NO. AFFECTED: D - 1901	DWG TITLE: Actuator Mounting Clevis

REASON FOR CHANGE:

    Add group to accommodate new actuator size

DESCRIPTION OF REQUESTED CHANGE:

    Add group 3 to dwg. Same as group 1 except change the following

dimensions for group 3

        Hole dia to be 1.249 - 1.251 in.
        A dim. to be .745 - .755

        Ref: Actuator Dwg. G-2171

CHARGE NO. 9188	CHANGE
SPECIAL REQUIREMENTS: None	**REQUESTED BY:** *Ron Joseph*
	APPROVED: *J Campbell*
	DATE RECEIVED: **SHEET 1 OF** 1

FIGURE 3.12 Drawing change request form.

design team leader or engineering design supervisor. It may not be
needed for control of detail changes; however, it may be beneficial
if the originating engineer is relatively junior in experience, or if the
product is complex, and an erroneous change can have severe consequences. More on this is covered in Chapter 5.

3.6.2 Revisions to Drawings

There are several practices in the actual revision of drawings that
contribute to proper control and provide traceability of the design
changes. The first of these is the description of the change on the
drawing itself. For many years, it has been a standard practice to
list what has been changed on the drawings in a revision column along
one edge of the drawing border line. In many cases, this is followed
faithfully and provides a clear and understandable record of what was
changed. Each time the drawing is revised, a new set of entries are
made in the revision column, and each revision is identified by the
next sequential letter or number as shown in Figure 3.13. However,

SUB	REVISIONS
1	ADDED NOTES 3-5 .868 WAS .848 1.25 RAD. WAS 1.00 ADDED 8.75 DIM. *C. Thornburg* *1/14/82* *Steven Paul* *1/14/82*
2	Changed Spec M-122 To M-149 *a. Kemp* *11/3/82* *J. Warren* *11/3/82*

FIGURE 3.13 Drawing revision column.

as many engineers have recognized, the draftsman is inclined to describe the change in the briefest of terms. In fact, it is common to find that only the most obvious elements of the change have been listed at all.

An effective way of overcoming this loss of detailed description is to require that each change request document be uniquely numbered for identification. Then when the affected drawing is revised according to the drawing change document, the change document number is listed in the drawing revision column. This approach provides traceability to the change, maintains a detailed record of the change, and makes it easy to trace whether or not a specific change document has been incorporated into the drawing. Although this method sounds deceivingly simple, it is not used in nearly as many engineering departments as it should be.

To provide the proper backup of the changes, the drafting department is expected to maintain a file of the drawing change documents. One approach is to microfilm the changed documents periodically in batches. The files typically are arranged in numerical sequence by change document number to make it easy to find a particular change.

Another tool that is used effectively in large engineering departments is to cross reference every change document against every affected drawing number in the engineering computer. This gives a cross-check to the engineer and draftsman of what changes should be incorporated on each drawing.

One other element of design assurance can be applied to drawing revisions. It is a good practice to require the cognizant engineer or drafting supervisor to review and sign off each drawing after it is revised to verify that the drawing was changed correctly. This reduces the possibility of incorrect interpretation of the change or errors of omission. However, it is important to maintain integrity in the drawing revision process. Do not allow changes other than those defined on the drawing change document to be made—even if it is to correct an obvious error. Require that another change document be processed, especially if the product is in production or in customer use. Otherwise, many surprises, in terms of incomplete, incorrect or untimely changes, can raise havoc in the design control process. Be faithful to a disciplined and orderly revision process, and it will pay off handsomely in reducing errors and avoiding costly problems in the factory and the field.

There is one other aspect of drawing signoff to consider. With the advent of computer-aided-drafting (CAD), it is not possible to have handwritten signatures in the computer. However, since many engineering organizations use microfilm for drawing production, the print which is used as the master for microfilming can be signed by the drafting or engineering personnel prior to filming. Regardless of whether machine-typed or longhand signatures are used, the important factor is to maintain integrity in the system. The appearance of the name on the drawing or in the revision column must truly represent review and acceptance of the drawing or its revision.

3.7 SUMMARY

Drawings are the tools and the products of an engineering department. Many other groups must rely on the drawings in the accomplishment of their responsibilities. Whether the drawings are being used for purchasing, manufacturing, installation, maintenance or whatever, the users of the drawings routinely assume the documents are correct and perform their actions accordingly. Since the drawings have such a far-reaching impact on the performance of the product, its producers and its users, a little extra effort in the preparation, release and control of drawings by the engineering organization is truly warranted.

4

Specification Control

4.1 Introduction   63

4.2 Types of Specifications   64

    4.2.1 Material Specifications   64
    4.2.2 Process Specifications   65
    4.2.3 Product Specifications   69

4.3 Specification Preparation   72

4.4 Specification Review and Approval   76

    4.4.1 Material Specifications   76
    4.4.2 Process Specifications   76
    4.4.3 Product Specifications   77
    4.4.4 Specification Signoff Practices   77

4.5 Specification Document Control   78

    4.5.1 Specification Numbering Systems   78
    4.5.2 Specification Release   79
    4.5.3 Specification Distribution   81

4.6 Specification Revisions   81

4.7 Summary   83

4.1 INTRODUCTION

The technical specification is another one of the basic tools of an engineering organization. In many ways, it is like a drawing. A technical specification defines requirements for others to meet. However, it uses words instead of pictures to describe a material, a process or even a product.

Specifications are used in many industries for many applications. As a result, the size, style and format of specifications varies widely. For example, a material specification might be as small as a 3 × 5 inch card. In contrast, a specification for a nuclear powerplant might consist of hundreds of pages of text, charts and diagrams. Nevertheless, the basic principles for assuring the proper use and control of specifications apply to both extremes and virtually all cases in between.

This chapter will tend to parallel Chapter 3, Drawing Control, since many of the design assurance techniques applicable to drawings also apply to specifications. Chapter 4 describes the types of specifications most commonly found in industry and presents methods for preparing, approving and controlling them. It also addresses the preferred practices for revising specifications and controlling their distribution and use.

4.2 TYPES OF SPECIFICATIONS

In industry today, many different kinds of specifications are used. However, the vast majority tend to be used for one of three purposes:

To define a material which has a particular composition and set of properties.

To describe a process which will produce a specific condition or set of physical characteristics.

To define the functional requirements that a component or system must meet.

Each of these specification types are described below.

4.2.1 Material Specifications

Material specifications are in wide use in nearly all industries. These specifications define the requirements of the many and various raw materials and feedstock.

Material specifications may be used in any one of several different places. It is quite common to define the specifications for materials of construction on drawings or in a bill of materials. Process specifications frequently call out particular materials or ingredients to be used in the process. Likewise, materials specifications often are included in an equipment specification to require the use of only those materials listed.

A material specification is naturally narrow in its scope. It must focus on those particular properties and characteristics which are important and unique to that substance. Thus, it must be quite specific in the definition of requirements, if it is to be sufficiently discriminating between those materials that have the desired behavior and those that do not. As it turns out in practice, there are many instances where several variations of a basic material, or even several different materials, can be used satisfactorily in a given application. However, there are many other cases where only one material will perform properly. It is the task of the specification writer to define what is needed and do it in terms the users can understand.

A well-written material specification has several important characteristics. These include:

Identification: Provide a generic, or family-tree type of identification of the basic material; such as, a free-machining derivative of austenitic stainless steel; or, a highly-refined grade of selenium suitable for electronic components applications.

Key Properties: Define the important chemical and/or physical properties in terms of toleranced numerical values. These properties may be in tabular or graphical form.

Acceptance Criteria: Include a statement of how the material supplier must demonstrate compliance with the specified requirements. For example, this might be by furnishing test data, test reports, material samples, test pieces or certificates of compliance.

Packaging: Specify how the material must be packaged or protected. Also any special restrictions or hazards that apply to the shipment or storage of the material must be identified.

Marking: Define how the material must be identified and marked to assure proper recognition and control at time of delivery and in subsequent storage and use.

Figure 4.1 is an example of a simple material specification. This format presents the requirements in an easy-to-understand arrangement and provides tolerances and limits for the chemical and physical requirements. Although some specifications may consist of several pages of requirements, they still must address the same general elements illustrated here for proper design control purposes.

Another form of material identification is the Material Sheet (or M-Card). This can be used to simplify the identification of a generic material which otherwise would require many words to distinguish between its variations. It is also used in applications where the detailed requirements are not known, but experience shows that a particular brand name or a product from a known source of supply fulfills the requirements. The Material Sheet or M-Card provides a simple means of identifying the material in terms that are convenient to use on drawings, specifications or instructions. Figure 4.2 is an example of a Material Card.

4.2.2 Process Specifications

A process specification defines those parameters which must be controlled to obtain a desired outcome, such as, the hardness of a

CORTLAND MACHINE COMPANY
Los Angeles, CA

Material Specification 106794   May 5, 1979
Revision B

HOT ROLLED STEEL SHEET AND STRIP

1. This specification covers hot rolled, low carbon steel sheet and strip.

Designation	Description
106794-1	Pickled and oiled sheet and strip.
106794-2	Pickled sheet and strip, not oiled.

NOTE: UNLESS OTHERWISE SPECIFIED, THE FOLLOWING REQUIRE-
MENTS APPLY TO ALL GRADES.

2. No change shall be made in the quality of successive shipments of material furnished under this specification without first obtaining the approval of the purchaser.

Manufacture

3. Process: Open hearth, basic oxygen, or any other process approved by the purchaser.

4. Discard: Sufficient from each ingot to insure freedom from pipe and undue segregation.

5. Quality: "Drawing Quality."

6. Condition:(-1) Hot rolled, pickled free from scale, and oiled to prevent corrosion. (-2) Hot rolled and pickled free from scale, and not oiled.

Chemical Properties and Tests

7. Chemical Composition: % heat analysis

Carbon, max.	0.15
Manganese	0.25-0.50
Phosphorus, max.	0.04
Sulfur, max.	0.04

Physical Properties and Tests

8. Temper: Suitable for the production of miscellaneous deep drawn parts. The manufacturer shall assume responsibility for selection of steel, control of processing, and performance of the material within properly established breakage limits.

FIGURE 4.1 Material specification.

M-101-F   Aug. 5, 1975

STEEL STRIP, CARBON, SPRING, COLD ROLLED, (SAE 1050)
SUPPLIERS:

(A) Inland Steel Corp, Chicago, IL

(B) U.S. Steel, Pittsburgh, PA

(C) Jones & Laughlin Steel Co., Pittsburgh, PA

ORDER FROM SUPPLIERS AS - Strip, stating Spec Number and Rev.
Letter.

Characteristics: Cold rolled, untempered, spring steel strip:

Grade	Description
Hard rolled	Rockwell C hardness 23-29

Chemical Composition: %		
	Carbon	0.46-0.55
	Manganese	0.60-0.90
	Phosphorus, Max.	0.040
	Sulfur, Max.	0.050
	Silicon	0.15-0.30

Thickness, in.	Sheet Size, in.
Less than 0.025	6 × 36
0.025 to 0.040 Excl.	10 × 80
0.040 and over	20 × 120

Tolerances: See Corp Std 101T27.

Application: Carriers on tumbler switches; hardened parts.

Specify by: Spec No., Thickness & sheet size

FIGURE 4.2 Material card.

material, the deposition of a conducting film, or the joining of two
pieces. One special type of process specification, the test specification, is described later in Chapter 8, Engineering Tests.

Process specifications often tend to be "how to" documents.
In some companies, process specifications resemble cook books. The
specifications provide a step-by-step, operation-by-operation list of
actions to be taken. This approach may be workable for in-house
processes, because it is tailored to existing skills and available
equipment. However, the cookbook-style process specifications do
not work well when the process must be performed by a supplier or
subcontractor. The outside sources rarely ever have the same facilities and production capabilities as the originator of the specification.

AJAX MANUFACTURING COMPANY
HEAT TRANSFER DIVISION
Middletown, Texas

Process Specification P-257   Initial Issue: June 27, 1979
                                     Revision E: April 17, 1981

CLEANING OF HEAT EXCHANGERS

Scope and Application

This specification defines the procedures for cleaning newly-manufactured and refurbished heat exchangers in the factory.

Critical Materials and Equipment

Material	Ident.
Detergent	M-1653
Rust Inhibitor	M-1289
Heat Exchanger Cleaning Rack	Dwg. 184D625

Process Operations

1. Preparation of Cleaning Solution

    a. Fill reservoir to "Full" mark with tap water.
    b. Add 20 pounds of detergent (Material Spec. M-1653) and 10 pounds of rust inhibitor (Material Spec. M-1289).
    c. Turn on water heater and heat wash water to 160-180°F.

2. Cleaning on Newly Manufactured Heat Exchanges

    a. Install heat changer in horizontal attitude and connect water supply hose to primary side inlet.
    b. Connect return hose to primary side outlet.
    c. Open reservoir circulating valve and set motor timer for 5 minutes.
    d. Turn on cleaning rack pump and let wash solution circulate until timer stops pump.
    e. Elevate rack to vertical position to drain out cleaning solution.
    f. Return rack to horizontal position.
    g. Close reservoir circulating valve.

FIGURE 4.3 Process specification.

Thus, it is preferable to write functional specifications to control a process. The specification needs to define what requirements must be met; not how. From a design assurance standpoint, the process specification must specify the key parameters and the associated tolerance or limits. To the maximum extent possible, it should not limit the user to an exact sequence of events or to a particular piece of equipment.

The important ingredients of a process specification are summarized below.

Identification: Give the specification a name that is descriptive but limited to a few words. Recognize that long names will be shortened in a daily use anyway.

Scope and Purpose: Describe what the process covers and how it will be used. This is a snapshot for familiarizing the user with what to expect.

Applicable Materials: Define specifically those materials to be used that are important to achieve the proper results from the process. All such materials should be described by name and type. Wherever possible, list the applicable specification number for each key material.

Required Results: Define what output should be achieved from the process. This frequently is the acceptance criteria for the items which have been processed according to the specification. For example, these may be physical properties, such as, tensile strength or hardness; or functional characteristics, such as, plating thickness or freedom from delaminations.

Process Limits: It is common to specify limits on process parameters to control the final results. These may be the maximum, minimum or range limits on items, such as time, temperature, voltage, pressure, etc. Generally these limits are established in the development of the process. Exceeding these limits may cause undesired or substandard results.

Warnings and Precautions: It is frequently necessary to point out parameters or process steps that may be crucial to the outcome, present hazards or cause adverse conditions.

From a design assurance point of view, the process specification is a means of achieving a portion of the design intent. The process specification directs the supplier or manufacturer along an established path toward fulfilling the design requirements. As such, it is in the best interests of the engineering department to be actively involved in the process specification system.

4.2.3 Product Specifications

A product specification defines the functional requirements that a component or system must meet. Product specifications can be used for

small and straightforward devices, such as a ball bearing, or for very
large and complex systems, such as a space vehicle.

To a large extent, a product specification is like a word drawing.
It is intended to define the needs the product must satisfy without
completely defining the product's size, shape and mode of operation.
Figure 4.4 illustrates a representative product specification.

For a product specification to be effective, it must delineate the
functional requirements that the equipment must satisfy. The specification needs to focus on what is needed and how the equipment is intended to be used. The supplier must be allowed some flexibility in
how the requirements will be met.

It is often a difficult juggling act for the specification writer to
present specific needs without severely restricting, or even dictating,
the supplier's design. However, this is the challenge. The contents
of the specification are very important to the final outcome of the product. If the specification is too loose or omits important design requirements, the resulting product may fail to perform satisfactorily in
its intended application. On the other hand, if the specification is
overly restrictive, it may prevent the suppliers from taking innovative
approaches for fulfilling the designer's requirements. In some extreme
instances, specifications might even contain contradictory requirements
which would result in confusion and lost time in the early stages of a
project.

To avoid these pitfalls, a few basic principles have evolved which
are important to consider from a design assurance standpoint. These
are summarized below.

Strive for clarity. Its importance cannot be overemphasized. The
specification must define what is wanted in clear and understandable terms.

Stick to basics. A specification is a requirements document. The
writer must distinguish between requirements and wishes. If it
isn't really needed, don't include it.

Define the acceptance criteria for requirements in terms of specific,
measurable values and appropriate units of measure. Give either
tolerances or limits for each quantified requirement.

Minimize the use of requirements which cannot be quantified or
verified by some method of analysis, inspection or test. Recognize
the difficulty of demonstrating compliance with an unmeasurable
requirement.

Pay careful attention to the definition of the interfaces with other
components or systems. Describe the interfaces fully. Include
diagrams, as needed, for clarify.

Where only certain parts or materials are permitted for use, define
such items in specific terms so the item and its source of supply
can be located.

QUALITY SYSTEMS CORPORATION
Philadelphia, PA

SINGLE-STAGE CENTRIFUGAL PUMPS   Spec. No. GTH-1814
                                      February 1, 1980

Scope

This specification defines the requirements for single-stage, centrifugal pumps to be used in irrigation and water supply systems.

Applicable Documents

American Voluntary Standard for Centrifugal Pumps, October 1975

Performance

1. Capacity
   The pumps shall be capable of pumping water at the following rates:
   Model   I: 0-5 gpm
   Model   II: 5-50 gpm
   Model   III: 50-150 gpm
2. Flow Characteristics
   Figures 1, 2, and 3 define the flow-head characteristics for Models I, II and III, respectively.
3. Inlet Conditions
   The pumps shall be free from cavitation at a minimum suction head of one foot.
4. Prime Mover
   Model I and II pumps shall be driven by 220 volt, 3 phase, squirrel cage motors.
   Model III pumps shall be driven by 440 volt, 3 phase, squirrel cage motors.
5. Reliability
   The pumps shall be capable of running on a continuous duty cycle without loss of pump delivery for 8000 hours.
6. Maintainability
   The pumps and motor shall be assembled and packaged in such a manner to permit the removal and installation of the complete assembly with commonly-available tools and equipment in the time periods listed below.

Model	Time to Replace
I	one hour
II	two hours
III	four hours

FIGURE 4.4 Product specification.

Group related requirements and use an abundance of sub-headings
in the specification to identify the various categories of requirements. These serve as road-signs and help the user understand
the content of the document.

Product specifications are very useful documents. However, engineering departments must realize that their specifications become legal
documents when applied by contract or purchase order between a buyer
and a seller. Specifications can frequently be the cause for disputes
and financial claims. Consequently, engineering organizations need to
exercise thoroughness and care in the preparation and use of product
specifications. It is another important building block in assuring the
quality of the product design.

4.3 SPECIFICATION PREPARATION

In most companies, specification preparation tends to be less formal
than drawing preparation. It is generally less disciplined and largely
left to a small group, or perhaps one person, to control. There is very
little in the industrial literature on the subject, and there are no widely-
accepted standard practices as there are in engineering graphics.
Therefore, each company tailors its specifications to its own perceived
needs. Figure 4.5 describes a typical specification cycle.

The task of preparing a specification is generally assigned to one
individual. Small companies delegate it to whoever is most knowledgeable (or most available). Medium-sized companies may assign the duty
to a specialty group that has some expertise in the technical subject.
Large firms, especially those in advanced technology industries, often
have specification writers that gather the technical inputs from specialists and compile the information into a pre-determined format. They
control the entire specification production and administration function.

The primary tasks in specification preparation are to decide what
requirements apply and develop a suitable format for presenting the
requirements in a clear and logical manner. Determining what is
needed is generally the most difficult for the inexperienced specification writer. Arranging the requirements usually will follow once the
requirements for the various topics are established.

To this end, suggested topic outlines are presented in Table 4.1
for material specifications and in Table 4.2 for process specifications.
These listings are general in nature but will fit most circumstances.

Product specifications may cover a wide range of apparatus, and it
is difficult to generalize what should be included. However, the Air
Force Systems Command developed a standardized specification format
that the United States Air Force has applied to many types of equipment and systems. The format has been successfully adapted to many
other applications and may be useful as a guide to specification writers

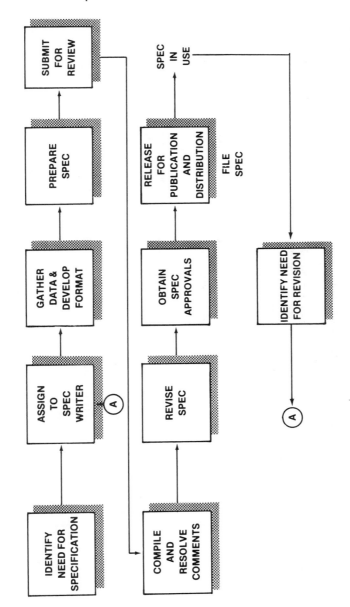

FIGURE 4.5 Typical specification cycle.

TABLE 4.1 Typical Contents of Material Specifications

1. Identification	List applicable specification title and number.
2. Scope and application	Provide a brief description of the generic or general type of material covered by the specification. Also briefly state how the material will be used.
3. Method of manufacture	Describe allowable processes which can be used.
4. Chemical properties and tests	Define the important elements and values for the chemical composition. Specify any tests which must be performed by the material supplier.
5. Physical properties and tests	Specify those mechanical and physical properties that must be met. Define any tests that must be performed by the material supplier.
6. Dimension	State sizes/shapes to be furnished and applicable tolerances.
7. Finish	Define special conditions of the material, as applicable.
8. Packing and marking	Specify how the material must be packaged and marked for shipment.

TABLE 4.2 Typical Contents of Process Specifications

1. Identification	List applicable specification title and number.
2. Scope and application	Provide a brief description of the process and its intended use.
3. Critical materials and equipment	Define what materials must be used by name and specification number. List any special equipment that is important to the process.
4. Process details	Describe what is required, including preparation, sequencing and methods to be followed. (Use subheadings for clarity.)
5. Records	State what must be recorded, where, when and how.
6. Acceptance	Specify what inspections or tests must be performed and what criteria must be met to demonstrate that the process is acceptable.

in non-defense industries. It is presented in slightly abridged form
in Appendix 1.

Once the specification writer has developed a draft of the specification, it should then be subjected to an internal review and approval
cycle before the document is issued.

4.4 SPECIFICATION REVIEW AND APPROVAL

As mentioned earlier, there are very few rules or standard practices
published for the control of specification preparation. This applies to
the review and approval process as well. For design assurance purposes, several guidelines and preferred practices are presented in
this section. These are based on experience and observations of methods used in many different industries.

4.4.1 Material Specifications

By their very nature, material specifications are a special breed of
documents. The vast majority of technical knowledge and input for
this type comes from the metallurgists, chemists and physicists. Although other groups such as, design, manufacturing, quality control
etc., may participate in the review, the real expertise nearly always
rests with the materials specialists. Therefore, it is generally good
practice to circulate the specification for comment to those groups that
will be directly involved in its use, but give the final authority for
disposition of the comments and approval of the completed specification
to the materials specialists. However, even among specialists, the practice of two levels of review is usually beneficial. A second trained
person can provide an objective review and possibly avoid oversights
or critical errors.

In the event that a small company does not have anyone on its
staff that is sufficiently skilled to fulfill this role, the company can
seek outside assistance of a materials consultant. Such persons or
leads to them, can often be found by contacting a neighboring university, a local technical society, or a nearby larger manufacturer. The
consultant can assist in the preparation and/or the review of the specification. However, final responsibility for the specification must remain with the employing company. Nevertheless, the services of a
knowledgeable professional can be very helpful.

4.4.2 Process Specifications

Process specifications differ somewhat from material specifications.
Manufacturing processes, especially newly-developed ones, tend to
take on a local flavor. These processes are often strongly influenced
by available equipment and facilities. Therefore, process specifications

frequently are written around the company's experience, which may be considered proprietary information. This aspect contributes to the general lack of standardization throughout industry, even though a large share of "proprietary" processes are neither secret, nor unique.

Nevertheless, there are some guidelines that should be considered for the review and approval of process specifications. Most all processes are developed in a laboratory or in a pilot production line. Persons responsible for these activities usually are a part of the manufacturing, manufacturing engineering, or materials/processes development group. Regardless of its name or location within the organization, the group that develops the process and the group that must use the process should have major roles in the review and approval of the related process specifications. Representatives from these two groups should evaluate comments from other interested groups, such as design, materials engineering, quality control, etc. and make final disposition on the inputs and comments. Again, there is something favorable to be said for a second-level of approval from the group that has the most knowledge about the process.

4.4.3 Product Specifications

Product specifications are closely aligned to the design process. These specifications are very similar in purpose and origin to engineering drawings. Since they are used to define a component or system, the content is largely technical, and the majority of requirements usually come from the design engineering group. Although several other disciplines, such as, materials, manufacturing, marketing, quality/reliability/maintainability, safety, etc. may provide inputs for selected sections of the specification, design or systems engineering should play the strongest role in the disposition of comments and in the final approval of the completed product specification.

There is another aspect of requirements to be considered for these documents. Product specifications are generally used in the bidding and contract process with either customers or suppliers. Therefore, it is prudent to require all such specifications to be approved by not only the responsible design or equipment engineer but also by the chief engineer, the applicable project or systems manager, or the product line manager. In some cases, perhaps all of these are needed in major projects where the stakes are high. This assures their involvement and commitment to the product, and they may contribute the wisdom of their own knowledge and experience early in the design cycle.

4.4.4 Specification Signoff Practices

The question of how to document specification approvals is often asked. One school of thought favors recording the actual approval signatures on a page of the final specification. The other approach is use a separate signoff sheet which is retained in the company's files but is not a visible part of the completed specification. Still others prefer to have a single approval signature, such as by the responsible engineer or department manager, appear on the cover sheet or on the last page of the specification.

Again, there are no hard and fast rules for this. However, experience shows that the majority of industrial firms seem to avoid displaying approval signatures on specifications which are issued for use external to the company. For specifications used exclusively internally, it is common to include the signoff sheet as part of the actual specification document. Some consider this evidence of commitment, others see it as strength in numbers (or misery loves company). From the design assurance standpoint, it really does not make any difference whether the signatures are a part of, or separate from, the specification. However, the involvement or two or more knowledgeable persons in the specification review and approval cycle is generally a valuable contribution towards achieving the best possible quality of the specification contents. Getting people to sign their name usually raises the level of their own personal commitment. The use of some form of specification signoff sheet, whether integral or separate, provides tangible evidence of participation in the approval process and, all things considered, is a good practice to follow.

4.5 SPECIFICATION DOCUMENT CONTROL

There are a few administrative aspects of specification control that do help the design assurance efforts. Although these elements are similar in principle to those used for controlling engineering drawings, the detailed techniques are different when applied to specifications. Methods for specification identification, release and distribution are presented in this section.

4.5.1 Specification Numbering Systems

Each specification needs its own unique identification by title and specification number. In addition, there needs to be some central control point for assigning and maintaining the specification numbers. Many different kinds of numbering systems are used in industry. Yet no one method appears to be dominant. It is the author's preference to use a simple, partially significant numbering system, consisting of a letter, designating the type of specification, followed by three or four identifier digits and a specification revision letter. Such a numbering scheme is illustrated in Figure 4.6.

Nearly any type of numbering system is workable, if the following key design assurance principles are faithfully followed:

Material specification number:

M	164	A
Type of spec ("M" for Material)	Spec identifier number	Revision letter ("A" for original release)

Process specification number:

P	273	B
Type of spec ("P" for Process)	Spec identifier number	Revision letter ("B" for first revision)

Product specification number:

E	1299	D
Type of spec ("E" for Equipment)	Spec identifier number	Revision letter ("D" for third revision)

FIGURE 4.6 Specification numbering system.

Keep it simple—make it easy to use.
Develop a set of rules and apply them consistently.
Use it carefully to avoid errors and duplications of numbers.

4.5.2 Specification Release

The purpose of a specification release system is to tell the interested
parties that a particular specification is now available for use. The
system may be nothing more than distributing copies of the specification to those persons or groups who need it in their work. Another
system often found in industry is the specification index list. Periodically new or revised specifications are added to the list. Copies are
then sent to a pre-determined distribution. It is then up to those
persons having access to the list to use or apply the particular specifications of direct interest to them. The specifications may be grouped
on the list by type, project or application, in addition to a straight
numerical listing. An example of this is shown in Figure 4.7. This
method is often used by small or medium sized companies that do not
have a large number of active specifications (perhaps 100-200).
    Another release system is to use some form of specification release
document, such as an Engineering Order or a Document Release Memo-
randum. This document announces the availability of the new or

SPECIFICATION INDEX
June 30, 1982

Spec No.	Title	Revision letter
Material Specification		
M-1000	Cold rolled steel sheet	C
M-1001	Cold rolled steel plate	E
M-1002	Seamless steel pipe	A
M-1003	Aluminum sheet	B
M-1004	Hot rolled steel plate	C
M-1005	Carbon steel structural shapes	F
M-1006	Brass bar stock	C
M-1007	Aluminum rods and bars	D
M-1008	Copper plate and strip	B
M-1009	Phosphor-bronze spring strip	A
M-1010	Stainless steel sheet	B
M-1011	Stainless bar stock	B
Process Specifications		
P-050	Chromium plating	D
P-051	Cadmium plating	B
P-052	Painting of metal structures	J
P-053	Sand/Shot blasting	C
P-054	Welding of carbon steel	F
P-055	Welding stainless steel	K
P-056	Magnetic particle inspection	G
P-057	Copper brazing	E
P-058	Crimping electrical connectors	D

FIGURE 4.7   Specification index list.

revised specification by specification title, number and revision. The release document typically presents a one-paragraph abstract of what the specification covers and its intended application. The specification release document normally is signed and issued by the responsible engineering manager or by the specification writing group as applicable. Copies of the release document are usually distributed widely in large companies. On the other hand, if the specification only applies to a particular project or application, the release may be sent to the restricted distribution of persons who will need it in their work.

The list of specification numbers and titles is another package of information that may be included in the engineering computer files.

In this form, it is easy to update the list. Persons seeking information about the current revision status and applications of a particular specification can then interrogate the computer file. This is gaining increasing favor in organizations that use large numbers of specifications or that make many revisions to their specifications.

4.5.3 Specification Distribution

Specifications frequently do not receive as wide a distribution as drawings. As a result, the distribution system may be very simple. The originating group is often designated as the keeper of the specification, that is, the group that prepares and releases the specification retains the master copy and supplies copies, as needed, to users. This same group is responsible for maintaining the index and controlling the revisions to the specification.

In firms that use large numbers of specifications, a centralized document center may actually retain the released master and handle the reproduction and distribution of copies. However, control of the specification master list and the application of the various specifications generally remains with the engineering department.

For design assurance purposes, the most important aspect of specification distribution is the integrity of the system. It must make available the correct revision of the applicable specification on a timely basis to those persons who must use it. Anything less than that degrades the quality of the process and, possibly, the product.

4.6 SPECIFICATION REVISIONS

During the course of a specification's life, it will probably be revised several times. Methods for accomplishing and controlling revisions to specifications are not covered very thoroughly in the literature. Nevertheless, one might expect that the same approach used for drawings would be used for specifications. However, it doesn't necessarily work that way.

Revisions to drawings are often limited to relatively small changes to lines, dimensions and notes. In most cases, the original document is retained and modified, and a brief description of the changes is recorded in the revision column.

In contrast, changes to specifications frequently result in extensive changes to many sentences or paragraphs and occasionally affect several pages. Since specifications are normally typed documents, it is often easier to prepare one or several new pages to accomplish the revision, instead of revising the original sheets. Also, the changes are seldom described in the revised specification. Thus, it is not easy to determine what is different about a newly-revised specification.

In an attempt to overcome this difficulty, some companies place a mark or symbol in the margin, adjacent to the line or paragraph that was changed. However, this still is only a flag and can be clumsy to administer. Even with these markings, it is still necessary to compare the revised specification to the prior version to find all of the changes.

One simple method is to include a revision page in the specification, either immediately following the cover page or as the final page, which provides a brief narrative summary of the change. It at least conveys the intent and scope of the change so the reasonably-knowledgeable user can grasp the magnitude of the change and its impact. Some companies continue to add subsequent revision statements to the change page in a cumulative fashion. However, after two or three revisions, this practice becomes confusing. It is usually adequate to list only the most recent revision statement. If more information is needed, it can be obtained by reviewing prior revisions or by going directly to the group that prepares and controls the specification.

When a revision to a specification is made, the revised specification should be subjected to a review and approval process which is consistent with that used for the original release of the specification. Some companies allow specification revisions to short-cut the initial review and approval system, but this is not an acceptable practice for design assurance purposes.

One technique for providing the necessary review and appoval is to use a form of engineering change request to revise an issued specification. The change request defines the reason and scope of the change and may define changes in specific requirements. The document is circulated for review and approval to those persons or groups who signed off the original specification. After the engineering change request is approved, it serves as the authority document for the actual specification change. Upon completion, the document number of the change request is listed on a change control

---

CLEANING OF HEAT EXCHANGERS

Description of revision E

This revision changes the mixing instructions for preparing the cleaning agents and modifies the rinse time. An alternate drying cycle has been added.

Revision authorized per ECR10244.

FIGURE 4.8 Specification change control page.

page in the specification, such as illustrated in Figure 4.8. This approach provides clear traceability of the source and reason for the change to each revision letter of the specification. Such a detailed control process may not be required in all industries. However, it is simple to administer, and it does provide a useful record of specification changes.

Regardless of the method of revision used, each revision of a specification should carry its own revision identifier. Most frequently, this is done by assigning the next sequential revision letter. It is also a common practice to list the date of the new revision, either in place of the date of the prior revision or in addition to the prior date. The word "Revised" usually precedes the date to emphasize the change.

Some companies control the revision status of each individual page of the specification separately. This supposedly permits changing or retyping only those pages affected by the actual change. Revised pages then are assigned the next revision letter, and unchanged pages continue to carry the prior revision letter.

When this method is used, it is also necessary to include a revision status page which lists each page number and its applicable revision letter. This approach is most frequently used for product specifications, especially those with a large number of pages.

A more common practice is to just assign a new revision letter to the entire specification. This practice is often applied to material and process specifications, since they generally only consist of a few pages (often less than ten), and it easy to retype the entire specification. The expanding use of word processing equipment for typing and revising specifications makes the revision process much easier than in the past. This also automatically respaces and renumbers the pages as required by the extent and magnitude of the revision. It is then convenient to apply the same revision letter to all pages of the revised specification.

Additional details for controlling changes to specifications are presented in Chapter 5, Configuration Control.

4.7 SUMMARY

Specifications are another important tool for engineering departments to use. Specifications tend to be word drawings, providing a written description of a material, a process or a product. As such, these documents require elements of control which are similar to those applied to engineering drawings. Specifications define requirements, and the design assurance system must verify that the requirements are correct and see that the correct information is distributed to those persons who must use it.

5

Configuration Control

5.1 Introduction   84

5.2 Configuration Identification   86

    5.2.1 Bills of Material   86
    5.2.2 Parts Lists   87
    5.2.3 Controlled Assembly Parts Lists   89

5.3 Control of Changes   90

    5.3.1 Change Control Forms   90
    5.3.2 Classes of an Engineering Change   92
    5.3.3 Preparation of Engineering Changes   93
    5.3.4 Review and Approval of Changes   95
    5.3.5 Incorporation of Engineering Changes   96

5.4 Configuration Verification   102

5.5 Summary   103

5.1 INTRODUCTION

Simply stated, configuration control is the management of engineering change. In the previous chapters, the discussion focused on defining the requirements for the product and properly translating those requirements into the drawings and specifications. However, it is inevitable that in the lifetime of the product changes to the design will be necessary. Controlling this change process is a major element in assuring continuing quality in design. A change can be required for many different reasons, such as, in response to a customer request, the introduction of a performance improvement, or the correction of an error. Regardless of the reason, a controlled method is needed to see the revision is made correctly.

Unfortunately, each engineering change represents an opportunity for something to go astray. Murphy's Law is especially applicable to the area of engineering changes—If anything can go wrong, it certainly will. Since each revision can have a different effect than all others, it requires an orderly approach to evaluate the impact of each change carefully. For example, what is the effect of the change on the strength of the component? How does this revision affect the inventory of existing parts? Does this revision require modification to the parts lists or instruction manuals? Do all existing units have to be changed or only those to be made in the future? These and similar questions need to be asked and answered during the consideration of a proposed change. Failure to do so can be both embarrassing and expensive.

Obviously the very simple components are less susceptible to revision control problems than are the complicated products. Nevertheless, it is important for the organization to examine each change thoroughly and take the necessary actions to implement it correctly.

It is almost uncanny on how frequently a company gets into trouble when incorporating what was believed to be "just a simple change." Don't underestimate the propensity for problems to appear as a result of a revision to the product design.

For ease of understanding, the change control process is presented in this chapter in three distinct phases. These are: (1) configuration identification, (2) control of changes and (3) configuration verification. Each of these areas represent an important aspect in the management of revisions. Figure 5.1 shows how these elements are inter-related. Methods and practices for administering a configuration control system in accordance with basic design assurance principles are described below.

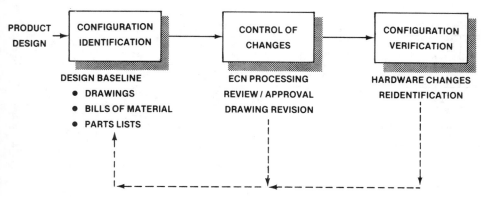

FIGURE 5.1   The change control cycle.

5.2   CONFIGURATION IDENTIFICATION

The starting point for control is the establishment of a design baseline. This can be the drawings, a bill of material or some form of
parts list.

The baseline represents the point of departure for change. As
each revision to the design is then incorporated in sequence, a new
design baseline is established. The process of orderly revision of
the drawings and associated bills of material or parts lists is the
foundation for configuration control.

5.2.1   Bills of Material

It is common practice to construct a list of parts and materials on
assembly drawings, as illustrated in Chapter 3. This compilation
is usually called the Bill of Material. It defines each item needed by
name; part number, specification number or similar identifier; and
the quantity required in that assembly. It also normally contains a
line item number (or "find" number) which is used to show where
that part or material is used in the assembly. This is illustrated in
Figure 5.2.

Some companies prepare the Bill of Material on a page separate
from the drawing to permit typing the information, rather than
printing it by hand as it had been done previously for many years.
More and more companies are now preparing and printing the Bill
of Material by computer. Nevertheless, the concept for control is
the same.

When a design change is made, the new drawing or specification number or quantity change is incorporated into the Bill of Material. This then establishes the new design configuration. For design control purposes, some notation of this change should be made

QTY REQ'D				CODE	ITEM NO.	DESCRIPTION OF ITEM	PART NO. OR IDENT. NO.	MATERIAL
G04	G03	G02	G01					
			1		1	ROTOR ASSY, TURBINE	64J1482G01	
			1		2	SUPPORT SHAFT, FRONT	357C191H01	CRES STL TYPE 416
			1		3	SUPPORT SHAFT, REAR	358C883H03	CRES STL TYPE 416
			1		4	DISK ASSY, STAGE 1	64J3155G02	
			1		5	DISK ASSY, STAGE 2	64J3174G01	
			16		6	BOLT, ROTOR	757B692H02	CRES STL TYPE 416
			16		7	LOCKING NUT	555C146H01	STL 4160

FIGURE 5.2   A typical bill of material on an assembly drawing.

in the drawing revision column. More on this is described in Section 5.3

5.2.2 Parts Lists

For relatively simple components and assemblies, the Bill of Material approach is adequate for control purposes. However, major systems, with perhaps hundreds or thousands of parts, are too large and too complex to control strictly by the Bills of Material.

The most common control technique is to compile a system parts list on a computer. It normally is structured to contain all of the parts, subassemblies and major assemblies in one large list. The parts list is routinely constructed in a manner to show each individual part that is contained in each subassembly and to show where each subassembly is contained the next higher level of assembly. Such structuring is described as an indentured parts list (also called a "gozinta list - this part "goes-into" that assembly). Figure 5.3 shows an example of a parts list with each subsequent level of assembly numbered to show that it is a part of the next higher assembly. A letter or number designator is often used to show the various levels of assembly, with "A" or "1" being the highest level and progressing downward to the individual parts or materials.

PARTS LIST 417395 REV. NO. 4

Level	Item no.	Description	Ident. no.	Qty.
A	1.0000	Turbine assembly, model 615	57J1822G01	Ref.
B	1.1000	Turbine rotor assy	64J1482G01	1
C	1.1100	Support shaft, front	357C191H01	1
C	1.1200	Support shaft, rear	358C883H03	1
C	1.1300	Disk assy, stage 1	64J3155G02	1
D	1.1310	Disk	459D662H01	1
D	1.1320	Blade, stage 1	334C129H01	24
D	1.1330	Retainer	781B375H02	24
C	1.1400	Disk assy, stage 2	64J3174G01	1
D	1.1410	Disk	459D663H01	1
D	1.1420	Blade, stage 2	334C134H01	32
D	1.1430	Retainer	781B375H02	32
C	1.1500	Bolt, rotor	757B692H02	16
C	1.1600	Locking nut	555C146H01	16
B	1.2000	Turbine stator assy	59J6252G01	1
C	1.2100	Casing, stator	63J7733G01	1

FIGURE 5.3 An example of an indentured parts list.

REVISION CONTROL SHEET PARTS LIST NO. 417395

Rev. no.	Item no.	Items affected description	Part no.
1	1.1200	Suppt. shaft, rear	358C883H01
2	1.2144	Bolt, flange	755B166H01
	1.2146	Bolt, flange	755B173H02
	1.2148	Bolt, flange	755B175H01
3	1.1200	Suppt. shaft, rear	358C883H02
4	1.1330	Retainer	781B375H01
	1.1430	Retainer	781B375H01

FIGURE 5.4 One type of revision control sheet for a parts list.

For design assurance purposes, it is usually necessary to assign the parts list its own unique identifying number. Then as each revision is incorporated, the parts list is reidentified by adding a revision letter or change number to the parts list number. Also some means of listing what changes were included in each revision of the parts list is very desirable. One approach is simply to list the engineering drawing numbers that were revised as a result of the parts list revision as shown in Figure 5.4. Another method is to list the document numbers of the engineering change documents that were incorporated in the parts list revision.

It is not hard to recognize that maintaining an accurate listing of all parts, materials and quantities is quite a challenge for complex equipment or systems. Frequently, one person, or a group of persons, must work fulltime to maintain the parts lists and introduce the approved changes. This may sound like an expensive operation and sometimes it is. However, failure to maintain a clear definition of the design can lead to loss of control and confusion in production, installation, operation and maintenance. That, too, can be very expensive. Loss of design definition also inhibits the engineering organization in trouble-shooting field problems, correcting product failures or introducing design improvements efficiently. Thus, investment of manpower in this effort is worthwhile.

The degree of control is a function of the number of parts, the frequency of change, the overall complexity of the item, and the seriousness of the consequences if loss of detailed design definition should occur. The greater the design complexity and the greater the seriousness of improper design definition generally requires a greater degree of control.

5.2.3 Controlled Assembly Parts Lists

In cases where the equipment or system contains a large number of drawings or parts, but the risk of design definition loss does not pose severe hazards or consequences, a reasonable level of design control can be achieved by focusing the control on the top levels of assembly. In effect, it applies tight control to a few selected drawings and key documents and relies on standard drawing/specification control practices to control individual details and minor assemblies. This is accomplished by a truncated parts list and is referred to as a Controlled Assembly Parts List (CAPL). Often it is prepared on an A-size drawing or 8-1/2 × 11-inch computer output sheet. Figure 5.5 illustrates one example of a CAPL to show content and format.

As changes to the drawings and assemblies are processed, two or more changes may be grouped and incorporated as a block. They are then treated as a package for both record control and physical modification of the hardware. An example of a CAPL change control page is presented in Figure 5.6. From this you can see that three engineering changes were incorporated in Revision A, and two changes were included in Revision B.

By identifying and listing the key elements of the design, there is a high probability that the most important changes will be the ones directly affecting those items or assemblies which are controlled. Thus, the necessary visibility and degree of control will be applied. Although it is not fool-proof, experience with the concept on large mechanical/electronic systems demonstrated it is a cost-effective

CONTROLLED ASSEMBLY PARTS LIST

Description	Ident. no.	Qty.
Turbine assy, model 615	57 J1822G01	1
Turbine rotor assy	64 J1482G01	1
Blade, stage 1	334C129H01	24
Blade, stage 2	334C134H01	32
Turbine stator assy	59 J625G01	1
Vane, stage 1	335C016H02	24
Vane, stage 2	335C021H02	32
Outline	123E474	Ref.
Wiring diagram	67 J8244-C	Ref.
Fire protection system	443D591	Ref.
Installation bulletin	IB615-2	Ref.
Instruction manual	T615C	Ref.
Lift bar assy	450D687G01	1

FIGURE 5.5 A controlled assembly parts list.

REVISION CONTROL SHEET CAPL NO. 1230
TURBINE ASSY, MODEL 615

Rev. no.	ECN no.	Items affected	New ident. no.
A	2122	Wiring diagram	67J8244-B
	2166	Vane, stage 2	335C021H02
	2185	Instruc. manual	T615C
B	2201	Wiring diagram	67J8244-C
	2217	Vane, stage 1	335C016H02

FIGURE 5.6   A CAPL change control page.

method of control.   Each firm must make their own decision on the
degree of required control.   The CAPL approach is worthy of consideration.

5.3   CONTROL OF CHANGES

The process of controlling engineering changes has many facets.
There are several administrative tasks to consider, as well as the
basic elements for preparing, reviewing and approving the technical
changes.
    Over the years, many different approaches have been developed
and used for controlling changes.   This section describes several
methods which have been used effectively in various design assurance programs.

5.3.1   Change Control Forms

The document used to define a proposed engineering change is an
important building block.   This document may be called an Engineering Change Request, Engineering Change Notice, Engineering Change
Proposal, Engineering Change Order, Revision Notice, or similar
title.   The purpose is to describe what is to be changed, why it is
necessary, and what are the consequences (impact) of making the
change.   In some circumstances, it may also be necessary to explain
what the consequences would be of not making the change.
    For purposes of explanation, the term Engineering Change No-
tice (ECN) will be used in this chapter.   However, the other titles
could just as easily be used.   Some organizations like to use differ-
ent-titled documents as the change moves through the complete cycle,
e.g. Change Request to Change Proposal to Change Order.   The

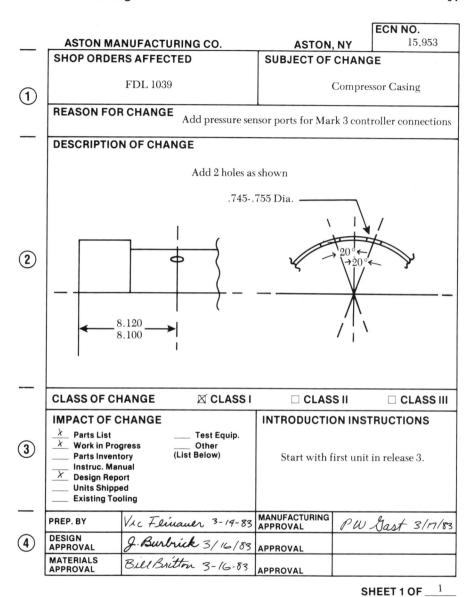

FIGURE 5.7 A typical engineering change notice form.

advocates of this approach claim it helps the personnel to understand the process. The critics of it say multiple pieces of paper confuses the situation more than it helps. In the long run, it is usually dealer's choice, although some government contracts may specify that certain change documents, by name, be used.

Figure 5.7 illustrates a typical ECN form. You will notice it is divided into four different sections. The top portion of the form contains the identification information, e.g., job number or project affected, the part or assembly involved, ECN serial number, and a brief statement of why the change is needed.

The next section provides space for the technical description of the change. For simple revisions, the block provided may be adequate. Otherwise, continuation sheets must be used.

The third section of the ECN form is used to highlight the impact of the change. A series of commonly-occurring effects are tabulated on the form and can be checked to show the change affects those items marked. Depending upon the complexity of the change or its impact on the system, it may be necessary to describe the effects in more detail on a continuation sheet or by attaching a memo to the ECN.

The bottom section of the ECN form is for the approval signatures. The ECN is not authorized for use until the necessary signatures are obtained. This aspect of the control of changes is discussed in Section 5.3.4.

5.3.2 Classes of Engineering Changes

Different schemes for categorizing engineering changes are used in industry. However, the system devised and used by the military services is the most common. The United States Armed Forces classify a change as a Class I or Class II. A Class I change normally is defined as affecting form, fit or function. This is further described as having an impact on contractual requirements, such as performance, physical interchangeability, reliability, safety, interfaces with other equipment, etc.

One other aspect of Class I changes must be mentioned. On government contracts, and on certain commercial contracts, the purchaser may impose a requirement to review and approve all Class I changes prior to implementation by the contractor. Of course, all internal approvals must be obtained before submitting it to the purchaser. This process adds a special logistics problem and routinely adds much time to the approval cycle. However, when required contractually, it must be accommodated.

In contrast, a Class II Change covers all other types of changes. It does not affect contractual requirements and can normally be approved and acted upon solely by the manufacturer. Nevertheless, it is still important to maintain control of Class II

changes. Although Class II changes do not affect interchangeability, it may still have to be recorded in the design baseline.

Some firms find it to their advantage to use a third category of
changes - Class III, for a Record Change. This class of change is
usually reserved for correcting obvious errors on documents. It is
restricted to changes which affect only the paperwork and not actual
hardware. Although this sounds simple enough, some changes can
be improperly classified as Class III and cause problems in control of
the actual hardware.

Another aspect of the change categories is the effect the change
has on hardware identification. A Class I change, by its very nature, implies that a potential user must be readily able to determine
whether or not the Class I change has been incorporated in the design. Class I changes routinely require that a new or revised part
number be assigned when the change is incorporated. In contrast,
a Class II change normally requires only that the drawing or specification be assigned the next revision letter or number when the
Class II change is incorporated. Likewise, a Class III, or Record,
change also only requires the correction be documented in the revision column or on the change control page.

The principle of reidentification is extremely important from a
design assurance standpoint. It provides objective evidence of the
revision to the baseline design configuration and provides traceability for the implementation of the change to the actual hardware.

5.3.3 Preparation of an Engineering Change

One of the first questions to be considered is: Who can initiate a
change? The most widely accepted answer is anyone can *request* a
change but only Engineering can *initiate* the official paperwork. The
purpose of this approach is twofold. First, it allows anyone in the
organization to call to Engineering's attention that a change appears
to be needed. Secondly, it attempts to prevent the formal paperwork
system from being swamped with change requests of uncertain validity. However, the engineering organization needs to establish a system for investigation and evaluating each request for change it receives. Some companies make it a policy of providing a reply to the
requester for each change request within a specified time period,
such as 10 working days. There is much to be said for that approach from a personal involvement/participation standpoint, but it is
not required for design assurance purposes. What is necessary,
however, is for the Engineering Department to recognize that someone believes there is a problem that needs to be corrected. Even
though their approach may not be the best, nevertheless, it is a
symptom of a problem or at least a misunderstanding. Engineering
should treat each request as potentially valid and investigate it.
Don't simply ignore it out of hand.

Once the engineer determines that an engineering change is necessary, it must then be documented for other persons to use. Several different techniques for preparing an Engineering change document are used in industry, including:

Preparation by the engineer or an assistant
Preparation by drafting
Preparation by a change control support group

Product complexity, quantity of change requests, and size of the organization are factors which influence the selection of the method to be used.

When the organization is small, and the product is simple, it may be adquate to have the engineer or a technician or draftsman prepare and distribute the Engineering Change Notice. However, as size, complexity and frequency tend to grow, it becomes increasingly necessary to establish and use a person or group that is dedicated to this task. Included in the scope of this job is the writing of the ECN and routing it for review and approval, the tracking of pending changes and the maintenance of the baseline definition of the product, typically the Bill of Material or Parts Lists. It is a vital task for maintaining effective control of the product design and is crucial in assuring quality in design.

High-technology programs, such as military and aerospace, and high-volume production programs, such as automotive or consumer electronics, often must develop and implement elaborate change control practices to manage the introduction of revisions to the product. This is frequently administered by technical support personnel, within the Drafting or Engineering Administration arm of the Engineering Department.

In these circumstances, the engineer initiates a draft ECN by filling out the form long hand, in enough detail so the administrative personnel can understand the basic intent of the change. The draft copy is then logged in and assigned a unique change control number.

Either significant or non-significant numbers can be used for identifying ECN's. There may be an advantage to use the job order number or project number as a prefix and successive dash numbers (e.g., -1, -2, -3) to identify individual ECN's. However, the most common practice is to assign a four- or five-digit, non-repeating number to each ECN as it is prepared (e.g., 2000, 2001, 2002). Then one single log can be used for controlling the assignment of all ECN numbers.

Next, the details of the change are investigated, and the final ECN is prepared. The package may contain sketches, word descriptions, dimensions, parts list revisions, etc. It should include a listing of specific drawings or specifications by number that are affected, a detailed definition of how the drawings or specifications must be changed, and a compilation of known effects the change will have in other areas, e.g., work-in-process, parts inventory, inter-changeability, design analysis, instruction manuals, etc. This in-vestigation can be tedious and time-consuming, but it is important for the decision maker to have the complete and factual data when evaluating a change. Accurate information improves the quality of decisions.

The reader may say, "that's all well and good if there is lots of time to do this, but what about the rush-rush situation?" By and large, the answer must be: "do what it takes to make a decision of acceptable level of quality (risk-benefit tradeoff)". From a design assurance standpoint, far too many sins are committed behind the cloak of expediency. Put another way - why is it we always have time to do it over, but we didn't have time to do it right the first time? A question well worth pondering.

After the ECN is prepared, it is then ready for evaluation and official approval for use.

5.3.4 Review and Approval of Changes

The review and approval process is much like the preparation proc-ess. It can be accomplished by several different methods.

The simplest is obviously just allowing the person that initiated the change approve it for use. Depending upon size, complexity, quantity, risk, etc. this may be perfectly acceptable. It certainly focuses the responsibility and accountability for the action. But when the objective is to enhance the quality of the design and the design decisions, two heads (or more) are frequently better than one.

There is a lot in favor of having each engineer's ECNs review-ed by his supervisor or by another competent and knowledgeable engineer in the organization. It is similar to the approval practices for drawings and specifications described in Chapters 3 and 4. This typically improves the quality of design decisions and often avoids errors due to oversights and misunderstandings.

As the product/process complexity and risk increases, there is a logical tendency to bring other specialists into the evaluation. Some companies ask several other departments to review and evalu-ate each engineering change. They then are expected to advise the appointed decision-maker (such as the engineer's supervisor, team leader or project manager) of the impact in their area of expertise. It is then up to the decision-maker to approve, disapprove, or send the ECN back for modification before final approval.

Another scheme is to route the ECN in daisy-chain fashion to each person who must approve it. This is generally done by hand-carrying the change to each person, discussing the pros and cons of the change, and finally obtaining the necessary approval signature. Sometimes a shortcoming is discovered somewhere in the review process, resulting in a modification that must be re-evaluated by persons who had already signed off earlier. Even though this approach seems cumbersome and inefficient, it is frequently found in industry. When applied with discipline, it can be effective for design assurance purposes.

The other method often found, and generally preferred, is the use of the Change Control Board (CCB). A Change Control Board is a group of appointed representatives from two or more disciplines. These persons have the authority and responsibility to act for their management in the evaluation and authorization of engineering changes. A CCB normally functions in a group meeting format. The engineer initiating the change or the CCB support personnel describes the need for the change and answers questions from CCB members about the change. A decision to approve or reject is normally made by the CCB chairman after hearing each member's report on the impact of the change. Other firms simply use the group meeting for discussion and then each person signs off the ECN. If the reviewers cannot agree on the change, it is referred to the next level of management or to the Project Manager for the final decision.

Scheduling of the CCB meetings is largely determined by the frquency and urgency of the situation. Major technical projects may hold CCB meetings daily or several times a week. Other programs may schedule the CCB meeting for once a week, or on demand as needed. Again, the local circumstances dictate and shape the system which is most suitable. Several variations of the CCB approach have worked well in achieving quality in design. Figure 5.8 shows an example of how an engineering change might be processed by a CCB.

An important principle to follow is to apply methods for reviewing and approving engineering changes which are consistent with the level of investigation and quality of design decisions that went into the original design. Anything less stands to degrade the design.

5.3.5 Incorporation of Engineering Changes

The introduction of approved engineering changes is frequently handled very informally. It is not unusual for the engineer to simply issue it without any advance notice or consultation. As might be expected, this approach can create havoc in the factory or in the field.

For design assurance purposes, discipline is needed to introduce the change in a planned and orderly fashion. Class I and Class II changes, by definition, have totally different impacts. These differences are summarized in Table 5.1. For instance, the

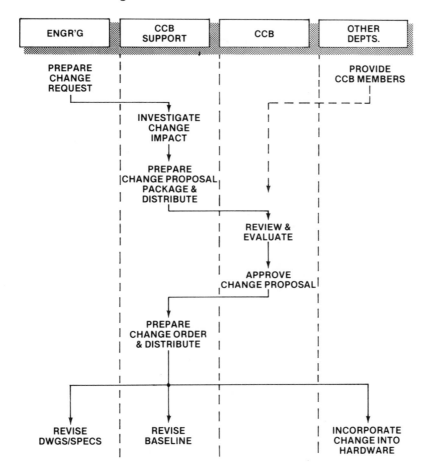

FIGURE 5.8   Flowchart for a Change Control Board.

actual introduction of a Class II change into the production cycle should be left to the manufacturing organization. This does not mean it should be ignored, but it should be incorporated at the earliest convenience or at the least-cost point.

On the other hand, a Class I change normally must be introduced as quickly as possible to minimize the impact of producing superseded designs. This requires a coordinated effort to make the change, since materials, tooling, completed parts, etc. are probably affected. As can be readily visualized, the orderly introduction of a Class I change cannot be left to chance. Even with full recognition of the impact of a Class I change, it requires careful orchestration to minimize the cost and disruption that typically goes with a Class I change.

TABLE 5.1 Class of Engineering Changes

Class	Design features affected	Interchangeability affected	Reidentification required	Approvals required	Hardware introduction
I	Yes	Yes	Yes	Internal and external	Must be defined and controlled
II	Yes	No	No	Internal	On basis of cost or convenience
III	No	No	No	Internal	N/A

Once the decision is made to authorize an engineering change, the information must be distributed to those groups that must act on, or be aware of, the revision to the design. This can be accomplished by providing revised drawings, specifications and related documents which incorporate the change or by providing a document, such as the signed-off ECN, for the people to use until the revisions can be incorporated into the affected documents.

There are two options available for incorporating the changes into the documents. The revisions can be made as part of the preparation of the ECN. In some situations, this presents a risk, since the review and evaluation process may result in further modification to the documents. The second option is to revise the drawings and other affected documents after the ECN is approved. However, this process puts the revision effort in series with making the design information available for use. Depending upon the local circumstances, this may be feasible. But in many instances, excessive delays are caused by this approach.

A more workable system is to release the ECN for direct use as soon as it is approved. Persons in the organization can use the information while the affected documents are being revised. Following this practice, though, requires a disciplined effort in the preparation and control of the ECN.

The first potential area of concern is with the actual wording of the revision. From a design control standpoint, no changes should be incorporated into a drawing or specification except as written exactly on the ECN. Therefore, accuracy and completeness of the change are important. Omissions or errors must be corrected by another ECN.

When a drawing is revised in accordance with an approved ECN, it is a recommended practice to list the ECN number in the revision block of the drawing as shown in Figure 5.9. This provides traceability to the change document and aids users of the drawing in determining whether or not the ECN has actually been incorporated.

If the Engineering Department policy is to define the change to the drawing completely on the ECN, it is then only necessary to list the ECN number in the drawing revision column instead of listing the details of the change. This, of course, necessitates keeping a copy of each new ECN in the Engineering record system for future reference.

The next area of concern is the bookkeeping for the design baseline. In many situations, it is absolutely mandatory to change the part numbers or modify the parts list for each change. This may require the pre-assigning of part numbers or document revision letters to maintain accurate baseline control, such as for equipment in production or items which must be modified, so all units will

REVISIONS				
SYM.	ZONE	DESCRIPTION	DATE	APPRV'D
A	D3	ADDED SECT. C-C	12/4/81	D Wilson
	A4	ADDED NOTE 5		JL Estes
		ECN 2675		
B	—	ADDED GROUP 2 & 3	4/6/82	D Wilson
	C8	16.75 WAS 16.57		JL Estes
		ECN 2761		
C	F7	GENERAL REVISION	9/27/82	T Smith
		ECN 2934		JL Estes

FIGURE 5.9   A sample of a drawing revision block, showing reference to the ECN number which initiated the change.

be the same. As can be imagined, this process is not easy to administer if several changes are processed against the same document in a short period of time. Close attention to detail is imperative.

Other circumstances may be such that only one or a few items are affected by a change. This may not require a one-for-one drawing revision for each ECN issued. Nevertheless, some method of accounting for the various changes is necessary. The block accounting for ECN's on CAPLS described in Section 5.2.3 is one possibility.

In the interim, many companies attach a copy of each approved ECN to controlled copies of the drawing. This system attempts to give visibility to the changes, but can be cumbersome to administer. Although there are no hard and fast rules, it is prudent to limit the number of ECN's in existence which have not been physically incorporated into an affected drawing or specification. Otherwise, the users will be confused by the changes and are likely to miss some aspect of a change. A rule of thumb is to limit the number of unincorporated ECN's to a maximum of about five. Depending upon the extensiveness of a particular change, the drawing may have to be changed sooner.

Another approach, which is less precise in the degree of control it provides, is the distribution or posting of a list of active changes by ECN number that have been authorized for a particular order or project. Figure 5.10 shows such a list. Generally, it is necessary to include date information showing when each ECN was released so a person can determine readily if any ECN's have been added since the last time they checked. Obviously, this process is less stringent

PROJECT ED-56 APPROVED ENGINEERING CHANGES

ECN no.	Date of release
1644	06-08-83
1647	06-10-83
1651	06-15-83
1652	06-15-83
1660	06-25-83
1661	06-26-83
1685	07-14-83
1693	07-30-83
1702	08-12-83
1739	09-21-83
1767	11-04-83

Authorized by:

Lee Touvelle

Project Manager

FIGURE 5.10   A project change control listing.

than the other methods described, but it is better than no control
at all.   If the listing is maintained in a computer file that can be
easily queried by persons with the need to know (e.g., engineering,
manufacturing, production control, quality control, etc.), it can be
effective for design control purposes.   A daily update of paper lists
displayed in a prominent location in the plant is another possible
approach, but it has obvious limitations from the control viewpoint.
     The guiding principle for design assurance is to get the revision
to the persons who need it in the performance of their jobs in a
timely fashion.   But, in addition, it is necessary to maintain a master list at a designated control point.   The master list serves as the
single reference if there are any questions about a particular change
or the status of the design.   Otherwise, errors or problems may
creep into the design documents, the hardware, or both.   When this
happens, it is often difficult to detect and correct it, unless a master list is maintained.   Even then, it may be difficult to recognize
the cause of the problem and take the proper action quickly.

5.4 CONFIGURATION VERIFICATION

After engineering changes are authorized for use, the next design
assurance task is to have a means of determining the status of pending changes and which changes have actually been incorporated in
the hardware. Again, equipment complexity, quantity and frequency
of changes, and size of the organization have a bearing on how easy
or difficult this task is. The verification task is downstream of engineering, but it is of interest to them. It is important to be able to
identify the hardware configuration clearly if other changes must be
made or a problem develops and must be investigated and resolved.

The key ingredient for configuration verification is appropriate
use of reidentification techniques. Wherever possible, it is preferable to have the change in identification marked on the product and
direct reading. This is typically accomplished by marking a new part
number or similar identifier on the equipment nameplate.

Another method is frequently used on very small items where
there is not enough physical space to mark a full part number. The
items can be color-coded, bar coded (as done on grocery store
items) or a code number or letter used to define the particular configuration.

Where there is a flurry of changes but configuration identification is required, it may be appropriate to list each ECN by number
on the components or on a special temporary nameplate as each revision is incorporated into the hardware. An example is illustrated
in Fig. 5.11. This situation frequently occurs in high-technology
systems where production is initiated in parallel with development.
Later, when the design definition stabilizes, a final nameplate with
the final part number can be added and temporary markings removed
or obliterated.

Engineering management must work with other departments in
the development of the system for configuration identification and

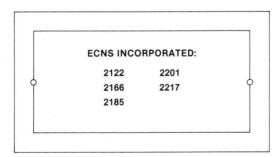

FIGURE 5.11 A temporary nameplate, showing ECN numbers which
have been incorporated.

verification.   As explained in this chapter, knowing the status of
ECN's in the system and the state of revision of the product is important for maintaining control and assuring quality in design.

5.5 SUMMARY

Control of engineering changes is an important part of any design
assurance program.   Each change presents an opportunity for a mis-
take to be made.   Thus, the processing of changes must be closely
controlled.

The three major elements of configuration control are configuration identification, change control and configuration verification.
Each of these elements involve a number of administrative details
which must be performed completely and accurately to maintain the
integrity of the design definition.   Thus, configuration control is
not a place to take shortcuts in any design assurance program.
Paying attention to the details really does make a difference.

6

Design Methods and Analysis

6.1 Introduction   104

6.2 Design Methods   105

    6.2.1 Engineering Procedures   106
    6.2.2 Standard Design Methods   113

6.3 Design Analysis   114

    6.3.1 Design Calculations   115
    6.3.2 Calculation Records   115
    6.3.3 Calculation Verification   118
    6.3.4 Technical Reports   120

6.4 Design Integration   122

6.5 Summary   122

6.1 INTRODUCTION

The design process is a combination of art and science. Developing a product which serves a useful function and packaging it in a manner which is both efficient and pleasing is an artform in itself. Yet scientific methods are required to assure the product makes effective use of materials, space, interactions among parts, and accomplishes this at a cost attractive to potential buyers. This aspect of design routinely requires mathematical analysis to determine the size and shape of parts to carry the required loads, operate for the prescribed life, withstand the environmental conditions, etc., in the course of fulfilling its intended functions. Some of the mathematical analysis may be simple, others may be very complex. Yet the principles for assuring the quality of design over the full range of extremes is essentially the same. Certain information should be defined, applied,

recorded and verified for all types of calculations. This chapter
discusses these basic elements and presents recommended practices
for implementation.

In addition to the mathematical aspects of design, there is another area where the orderly, logical approach of science can contribute to the design assurance process. That is in the development
and use of engineering procedures and standard practices. These
are tools for defining the preferred way for accomplishing the tasks
assigned to the engineering organization. In a one-man engineering
department the methods may be informal and simply committed to memory. But even still, the best methods based on the one engineer's
personal knowledge and experience will be used over and over
again—within the constraints of memory and the pressures imposed
by time and cost. Overall, each engineer typically wants to do the
job right and tends to stick to proven methods.

In the process of developing new products, engineers need to
apply scientific principles when considering new materials and design
concepts to understand how best to use these. This, too, needs to
be done in an orderly manner. Written procedures and standard
methods provide a planned and systematic method for performing the
work. They also allow less experienced engineers to benefit from
the years of experience and skill of those engineers who have pre-
ceded them. In addition, it offers the potential of having the best
methods used consistently. Consequently, a preplanned approach
makes a significant contribution towards assuring the quality of
design.

In the development and use of design procedures and standard
methods, each engineering organization must tailor those to its own
needs. The guiding principles here are: (1) Stick to basics; (2)
Keep it simple; and (3) Write it down.

6.2 DESIGN METHODS

Design methods describe the preferred ways that engineering management wants the work performed. In many companies, the methods
are passed along verbally through on-the-job exposure. This approach has been used for many years by many firms, with varying
degrees of success. However, its success is dependent upon the
significance of several factors. These include: the size and growth
rate of the company, its engineering staff, the number and complexity of the products and design methods; the skill and thoroughness
of the "trainers;" the relative stability of the company's and industry's technology; the consequences associated with errors, etc. In
some situations, this informal approach is quite satisfactory. In
others, it may be very inefficient, or perhaps even hazardous. Some
managers favor the "school of hard knocks" as the lasting training

experience. Yet, others refer to this as "trial-by-error." But the question is: Can you afford to allow the engineers to learn only from their own mistakes? In these times of dynamic markets, fierce competition, liability risks and rapid technological advances, successful companies long since concluded there has to be a better way. The "better way" routinely turns out to be an orderly process, built around proven and efficient methods and procedures, rigorously applied by all persons. The subsequent sections provide recommendations and guidelines for developing that orderly process for your engineering department.

6.2.1 Engineering Procedures

Many engineers complain profusely that all procedures and paperwork are just so much "red tape." They claim it is degrading and counter-productive. In many cases, they are correct and justified in their complaints. Complicated and unwieldy procedures, which stray from the basics, can make life far more difficult than necessary. But, procedures, records and methods which are fundamental to the core effort of engineering contribute to operational efficiency and product integrity.

The paragraphs below address the key elements that most engineering departments encounter and need to be prepared to accommodate in the daily work. From this, each organization can develop their own set of procedures to fit their needs.

Engineering Procedures: The first thing needed is to select the format and style of procedure. Determine whether or not a special page arrangement will be used, such as preprinted forms with company or department logo vs. a plain sheet. Each procedure should be individually titled, numbered, dated and assigned a revision letter for identification and control purposes. Table 6.1 lists the typical contents of an engineering procedure. Next, select the style of presentation you want to follow (simple running narrative paragraphs; play-script, flow diagram with notations; or some combination of these). Figures 6.1, 6.2, and 6.3 illustrate each of these styles. Define who can prepare a new or revised draft of a procedure and who must review and approve it. Specify how procedure changes can be accomplished. Decide where the approved procedures will be filed (procedures manual vs. central file vs. personal records, etc.) and tell the people where they can obtain copies for reference use.

Effective procedures also specify any forms to be used. This includes providing instructions for filling out the various blocks or spaces on the forms. Some companies find it helpful to attach a sample form that is filled out to show how the information is to be recorded.

TABLE 6.1 Typical Contents of an Engineering Procedure

Procedure number:	Assign a unique identifying number to each procedure. Include date and/or revision.
Title:	Define what the procedure covers (in about 3-6 words).
Scope/Purposes:	State what the procedure is intended to do and why it is needed.
Forms:	List the forms (by number) which are to be used in carrying out the procedure.
Procedure:	Define the steps to be taken and who is responsible for taking these actions. (Specify in functional terms, e.g., design engineer, design supervisor, engineering clerk, etc.) It may be desirable to divide the procedure section into titled subsections for clarity (e.g., Preparation of Change Request, Evaluation of Change, Incorporation of Approved Changes to Drawings/Specifications, etc.).
Attachments:	Include a copy of sample forms. Good practice to include instructions for filling out each form.
Approval:	Each procedure and later revision should show evidence of approval (e.g., company logo, approval signatures, typed statement of approval, etc.).

Design Requirements: Identify the common sources of the design requirements. Define how the designer is to obtain clarification of unclear requirements and how revisions to customer's specifications are to be handled. State how and where design requirements are to be filed, and describe how changes to original requirements are to be released for use in the design process. Refer to Chapter 2 for other factors to consider.

Drawings: Many engineering departments prepare and use a Drawing Room or Drafting Manual to define how drawings are to be made. These instructions and procedures normally apply to the drafter's work, such as line construction, views, dimensioning practices, bills of material, notes, etc. Separate procedures, in addition to those in the Drafting Manual, are often needed to define how a drawing is processed. For example, explain how to request a new or revised drawing and who must approve the request. Also once a drawing is

---

ENGINEERING PROCEDURE

Title: Drawing Changes

I.   *Purpose and Scope*
      To define how changes to engineering drawings are requested,
      evaluated and implemented.

II.   *Definitions*
      None

III. *Forms Used*
      Form No. 1612 - Drawing Change Request.

IV.   *Procedure*
      Anyone in the company can request a drawing change by pre-
      paring a Drawing Change Request (Form 1612). After com-
      pleting the form, obtain your supervisor's approval and submit
      the Request to the Drafting Supervisor.

      Upon receipt, the Drawing Change Request will be logged in by
      Drafting. The Change Request will then be submitted to the
      cognizant design engineer for evaluation. The design engineer
      will review the request and investigate the feasibility and valid-
      ity of the request.

      If the engineer does not concur with the need for the change,
      the engineer will explain why the change request is being de-
      nied. This will be written in the Evaluation block on the Draw-
      ing Change Request form. The evaluated Request will then be
      returned to the Drafting Supervisor. After logging out the
      Request, the Request form will be returned to the requester.

      If the engineer agrees the change should be made, the engi-
      neer will add any additional information or instructions to
      Drafting on the Request form and submit the Request form
      package to the Drafting supervisor. The Drafting Supervisor
      will then schedule the work and assign the Change Request to
      a specific Draftsman for action.

FIGURE 6.1 An example of the narrative style of procedure writing.

ENGINEERING PROCEDURE

Title: Drawing Changes

   I.   *Purpose and Scope*
        To define how changes to engineering drawings are requested,
        evaluated and implemented.

  II.   *Definitions*
        None

 III.   *Forms Used*
        Form No. 1612 - Drawing change request.

  IV.   *Procedure*

        A. Requesting a change:

            Originator   1. Prepare drawing change request
                                    (Form 1612).
            Originator's   2. Review and approve drawing change
            supervisor   request.
            Originator   3. Submit request to drafting supervisor
            Drafting   4. Log in change request.
            supervisor   5. Submit request to cognizant design
                                    engineer for evaluation.

        B. Evaluating drawing change request:

            Design   1. Evaluate feasibility and validity of
            engineer   request.
                                2. If disapproved:
                                    a. Write reason for rejection in evalua-
                                        tion block on change request form
                                        and return to drafting supervisor.
            Drafting   b. Log out and return change request
            supervisor   form to originator.
                                3. If approved:
            Design   a. Add any additional information or
            Engineer   instructions for drafting.

FIGURE 6.2 An example of the play script style of procedure writing.

ENGINEERING PROCEDURE

Title: Requesting and Evaluating Changes to Engineering Drawings

Originator

Prepare change request

Originator's supervisor

Review and approve

Originator

Submit request to drafting supervisor

Drafting supervisor

Log in and submit to
cognizant design engineer

Design engineer

Evaluate feasibility and
validity of request

Disapproved   *Approved*

Write reason on   Add necessary
request form   information or instructions

Drafting supervisor   *Drafting supervisor*

Log out and   Schedule work and
return to originator   assign to draftsman

FIGURE 6.3 An example of the flow diagram style of procedure
writing.

created, state who must review and approve it and how is it released for use. Procedures are typically required for specifying how released drawings can be revised and who must review and approve such changes. It may also be necessary to define special review and approval requirements for different products or projects, since a contract may introduce special requirements, such as project management or customer approval rights. Refer to Chapter 3 for additional factors to consider.

Specifications: The concepts for control of specifications are very similar to the control of drawings. However, separate procedures are normally required. The specification control procedures should define the requirements for preparation, review, approval and release of specifications and the control of revisions to released specifications. Individual procedures may be required, or at least tabulated separately, to cover the various types of specifications used, e.g., material, process, product, construction, etc. See Chapter 4 for additional factors to consider.

Engineering changes: Since an engineering change may have a far-reaching impact beyond just changing drawings or specifications, it is often necessary to prepare a separate procedure for the control of engineering changes. This is in addition to the procedures governing drawings and specification revisions. One example is the operation of the Change Control Board if one is used. Again, the procedure should define the forms to be used and who must be involved in the review and approval process. It may also have to provide direction for the release and implementation of approved changes, e.g., factory introduction point, field retrofit, etc. See Chapter 5 for additional factors to consider.

Calculations: Define how calculations are to be recorded. The procedure should specify the type of information to be included, requirements for checking the calculations, and where the calculations should be retained. See Section 6.3 for more details on the control of calculations.

Computer Programs: If the company uses a central computer or has a computer programming group, a procedure may be needed to define how engineers initiate a new program, run existing programs, and request corrections or modifications. If the engineers do their own programming and have direct access to operate computer terminals or personal computers, a procedure may be needed to define what verification steps are required and who else must be involved. See Section 6.3 and Chapter 7 for other factors to consider.

Testing: Define how engineers prepare and process test requests. Specify what approvals are required. Delineate the differences, if any, between various types of tests, e.g., development, proof, acceptance, reliability/life tests. Also define how test reports are to be prepared, approved and distributed. See Chapter 8 for additional factors to consider.

Design Reviews: Specify under what conditions design reviews are required. Define the basic responsibilities for conducting a design review. See Chapter 9 for other factors to consider.

Nonconformance Control: Disposition of nonconformances (factory, supplier, field, etc.) is frequently covered in a company or plant-level procedure. If not, the basic engineering responsibilities need to be defined in an engineering procedure. Some companies issue a one-page engineering procedure to state that the company or plant procedure is to be followed. Further details on controlling nonconformances are described in Chapter 11.

Engineering Records: Although other individual procedures may define what kinds of documents are to be retained and where they are to be filed, an overall records procedure is usually needed. It should define records retention periods and filing practices for the various types of Engineering records. For more information on this, see Chapter 12.

Technical Reports: Many engineering departments are required to write technical reports. Define the preferred report format and the process for review, approval and release of technical reports. Pay particular attention to the control of report distribution for patent and liability protection. See Section 6.3 for additional factors to consider.

Other: It is common to include certain administrative procedures in an engineering procedures system, covering topics such as time reporting, cost control, travel expenses, etc. Although these methods have little or no bearing on design assurance, there is a certain logic, orderliness and convenience for including these instructions in the engineering procedure system. This is, of course, a local decision.

Binder: For convenience, there is a lot of merit to file the approved procedures in a binder or notebook. It does not have to be fancy or multi-colored, but it should be easy to find and easy to use.

Index: Regardless of the method of retention, an index of procedures should be prepared and maintained. The index should list the procedure number, title and current revision (by number, letter and/or date of issue). It can be very similar to the type of index described for specification control purposes in Chapter 4. The index should be revised each time one or more new or revised procedures are issued, and a revised copy of the index placed in the procedures file and/or binders.

After an initial set of procedures are prepared and issued, the engineering organization needs to use the procedures for a reasonable trial period. Management should clarify, correct or streamline the procedures based on the results from the trial use. Once implemented, the procedures must be followed rigorously to get the benefit of using preferred practices consistently. However, the control process must be sufficiently flexible to allow subsequent improvements. Don't imply the procedures are "carved in stone" and can't ever be revised. Let the procedures work for you, not against you!

6.2.2 Standard Design Methods

Engineering procedures, discussed in the previous section, address the administrative aspects of design. Standard design methods address the technical aspects. Such methods represent the best thinking of the best engineers in performing the many and varied technical tasks. As the name implies, the standard design methods are the methods to be followed routinely. It is intended to limit the design practices on the recurring tasks to those methods which have been proven to work reliably. Such proof comes from extensive laboratory experimentation, comprehensive analysis studies or actual field experience (often from various combinations of these sources).

Standard design methods should be documented in a form much like an administrative procedure. Each design feature of importance should be described by a standard design method. In some cases where two or more conditions may be acceptable, each should be included. Table 6.2 presents a tabulation of ingredients that should typically be included in design standards documents. Additional detailed items may be necessary for specific applications. For example, a pump manufacturer might use standard design methods for: sizing the pump for capacity and head, selecting the type and size motor, choosing the proper shaft coupling, defining flange mounting, designing the pump rotor and casing for minimum losses, etc.

Each design standards document should be uniquely identified by title, document number and revision letter, number or date. For con-

TABLE 6.2 Typical Contents of a Standard Design Methods Document

Document number:	Assign unique identifying number to each Methods Document. Include provision for date and/or revision.
Title:	Define what the methods document covers.
General:	State the scope and purpose of the method. Provide brief background information about what it does and why it is needed.
Design method:	Specify how the technical task is to be performed (formulae; applicable criteria; design limits; tolerances to be used; applicable combinations of tasks, calculations, actions; acceptance criteria, etc.).
Notes:	List special instructions or limitations.
Approval:	Provide space for technical signoff (including date of approval).

venience it may be appropriate to use a numbering system which relates similar design features into a common section. For example, design standards for beam strength, column buckling, and torsional
loads could be grouped in a section on structural design. Design
standards for the design or selection of fans, pumps, and heat
exchangers might be grouped into a section on cooling equipment.

The standard design methods documents are normally supported
by either analysis or testing. Wherever possible, the source document,
such as particular calculation sheet or technical report number, should
be listed directly on the standard design document for easy reference.

Each standard design document needs to be reviewed and approved
by the technical experts of the company or plant prior to its release
for all other engineers to use. Each standard design document should
contain a technical signoff, associated with each revision. Some companies also require signoff by the responsible engineering manager in
addition to the technical specialist approval.

It is a good practice to compile all of the standard design methods
documents into a binder or manual. For a small department a single
copy may be all that is required. For the larger organization, several
copies can be located throughout the department, or it may be advis-
able to assign a copy to each engineer. However, the process of keeping each copy up-to-date takes on special importance to assure that
some persons are not working to obsolete technical instructions.

The use of a design manual (compilation of standard design methods) is a valuable tool for design assurance purposes. In addition to
enhancing the design process by getting all engineers to use the preferred methods consistently, it also helps the engineers recognize their
limits. The engineers are better able to realize when they encounter
an unusual situation which is beyond the boundaries of the standard
methods. At this point, they can get clarification or assistance from
the specialists and avoid possible problems in their designs.

6.3 DESIGN ANALYSIS

Since design analysis involves calculations of various types, it is important to set up an effective system for the control of calculations.
These controls can be used for both hand calculations and computer
calculations. The principles are the same, but the form and format
may differ.

This section describes the information that should be included in
the calculations, different means of recording and retaining the calculations, ways of performing checks of the calculations, and finally
a brief discussion of the preferred ways for controlling analysis reports.

6.3.1 Design Calculations

Many products today require different types of calculations to size the parts or to determine fits, motions, and loads that affect the safe and reliable operation of the product. As such, the calculations are an important basis for the design. Thus, the engineering calculations should be duly recorded and maintained as supporting records. Although many people joke about calculations being made on the back of an envelope, there are many occasions where that is close to the truth. All too frequently, the calculations are made on any available piece of paper, not clearly identified to the specific part or project, and haphazardly put away. It is frequently impossible to find the calculations, let alone check them for adequacy.

The problem is not restricted to hand calculations either. With the increased use of the computer in engineering analysis, it is easy to generate mountains of calculations in a short time. As a result, the analyst may lose track of the various computations and end up with unretrievable analysis.

To overcome these situations, greater discipline must be applied to the calculation process. Certain information is required as a minimum for design assurance purposes, and it applies to all types of calculations. These basic requirements are contained in Table 6.3. By applying these routinely to all calculations (whether hand or machine) much of the vagueness, uncertainty and lack of traceability can be avoided.

6.3.2 Calculation Records

The most common types of calculation records found in industry are loose sheets, bound record books and computer printouts. For hand calculations, the bound record books are desirable, since all sheets are pre-numbered, permanently attached to the record book, and are not easily lost. The disadvantage of this system is the difficulty of locating various calculations for different projects in record books. Another is the difficulty in finding an engineer's book when the engineer is not present. Some means of indexing multiple calculation books and calculations within each book is needed for retrieval purposes.

Loose sheets are preferred by many engineers because of the flexibility this method offers for using and filing. Yet it is this same flexibility that causes calculations to be misplaced or lost easily. Nevertheless, many companies favor making calculations on calculation forms or standard lined tablet sheets and then filing the calculations in related project and task folders.

Computer printouts present special problems of their own. The size and nature of the accordion-pleated pages cause handling and

TABLE 6.3 Minimum Requirements for Calculation Sheets

Purpose:	State the purpose and scope of the calculation (what is it and why is it being made).
Application:	Identify the program, project or product to which the calculation applies.
Method:	List the formulae used (or define the computer program name used).
Units of measure:	Define the units of the input and output data (psi, starting torque, $/unit, etc.).
Assumptions:	Define the key assumptions used in the calculations.
Results:	Show the results (with units) and define what the results represent (max. deflection, overload current, etc.).
Name and date:	Include the name or signature of the person performing the calculation and the date of calculation.
Identification:	Assign a file number or reference number to the calculation for retrieval purposes.

storage problems. Special binders are available, but they too are a size which does not fit in many desks and file cabinets. Also one printout looks like most other printouts. Microfilming is being used to reduce the size of the package, but it, too, has other problems of form and visibility. Again, a means of indexing and tagging is needed for retrieval purposes.

One approach for keeping track of calculations is to use a calculation summary sheet for each significant calculation, whether done by hand or machine. The Calculation Summary, illustrated in Figure 6.4, provides an overview of the purpose and scope of the calculation, key assumptions, the methods used, a summary of the results, and any special remarks or conclusions about the calculation. It also lists the calculation number of the computer printout or reference number assigned to hand calculations (e.g., page and calculation book number) for traceability to the actual calculations.

A few large firms involved in major development projects have incorporated a Calculation Summary system into their data storage and retrieval program. Each Calculation Summary is assigned a serial number and the title and document number are listed in a computerized data system, using key words for access and retrieval. In addition,

AJAX MANUFACTURING CO.	DOC. NO. 82-37
SUBJECT Flow and pressure drop for ADG Tower	**DEPT.** M.E.

1. DESIGN CONDITION/ASSUMPTIONS 2. METHODS USED 3. SUMMARY OF RESULTS 4. CONCLUSIONS

1. Design Conditions/Assumptions:
   a. Operating Temp.: $-20°F$ to $1250°F$
   b. Operating Press.: 300 psi max
   c. Steady State Conditions
   d. Use existing 2-in piping system

2. Methods
   Use FLOCALC computer program

3. Results:

Flow Rate	Press Drop
10 gpm	16.47 psi
20 gpm	40.60 psi
30 gpm	99.88 psi
40 gpm	233.20 psi

4. Conclusions
   2-in piping system is too small for use with ADG system which has design flow of 40 gpm.

PREPARED BY *Michael James* 10·6·82	APPROVED *n m Fullet* 10/6/82	SHEET 1 OF 1

FIGURE 6.4 Sample calculation summary sheet.

a copy of the Calculation Summary is microfilmed, and the film is serialized for indexing. Users can then search the system by selecting the project identifier or other key word descriptors to find the related document numbers of apparent interest. By viewing the microfilm for the indicated serial numbers, the engineer can decide which calculation is of interest, or possibly get the needed information directly from the microfilm display of the Calculation Summary.

6.3.3 Calculation Verification

Calculations routinely have been a personal action. In most companies, only the person who made the calculation ever sees it. Everyone else involved with the work assumes, rightly or wrongly, that the calculation is correct, accurate and appropriate to the situation. Unfortunately, that is not always true.

A basic process for assuring high levels of quality in calculations is to introduce some type of a second-party checking activity. The purpose of such a check is to minimize the chances for error or oversight by providing an independent and objective review. The verifier should be a person skilled in the same technology, knowledgeable of the application and have access to the applicable requirements, codes, standards and instructions. Further, the person performing this technical check should not be a person who was directly involved in the original work. This is intended to assure proper technical objectivity. In some situations, the engineer's immediate supervisor is excluded from being the reviewer, due to industry or contract restrictions. An ideal verifier would be a senior engineer who does similar design work on another project. In reality, it is often difficult, even impossible in small engineering departments, to find anyone who meets the preferred qualifications. In those cases, the engineer's supervisor may be the only person available with adequate technical knowledge to provide such a verification.

The primary purpose of reviewing calculations is to verify the adequacy of the analysis. This involves examining the appropriateness of the methods used, the proper coverage of the analysis, the validity of the assumptions, and the reasonableness of the results. Note that this does not always mean checking the accuracy, or mathematical precision, of the calculations. But in some situations, this may be required.

Experience shows it is quite common to find a computation was carried to many decimal places, implying great accuracy and precision. However, upon closer investigation, it becomes apparent that the assumptions are very imprecise, possibly even inaccurate. Consequently one of the most important aspects of an objective review of calculations is to examine the assumptions made by the analyst. This is where the judgment and knowledge of the independent reviewer comes into play, which explains why the reviewer needs to be an experienced engineer, rather than a junior engineer or someone of lower skills-level.

In regard to checking the accuracy of the results, such checks generally must be somewhat imprecise. Nevertheless, if the reviewer concludes the methods and assumptions used are reasonable, a correlation within 25-35 percent between the original results and the check made by a simplified method, is generally an acceptable level of verification. In some situations, you would expect to be closer. However, the simplified methods normally do not account for the effects of complex configurations, actual loadings or extremes in environmental conditions.

A fair rule of thumb would be to question any results which appear to be 50-75 percent different than the values the simplified methods give.

Verification of calculations can be accomplished in different ways. As noted above, one common technique for checking complicated calculations is to perform a simplified hand calculation, using well-established formulae. This method can verify the correctness of the original calculation in terms of order-of-magnitude, proper sign, and trend of the results. This provides a reasonable check and is an appropriate means for verifying new designs or for verifying work as an outside consultant might do.

Results from test programs or operational performance of similar designs are also used to verify the correctness of design analyses. However, caution must be used and steps taken to confirm that the item tested is sufficiently similar to the design being analyzed, so the results truly apply. When test results or operational data are used for verification, a record of the correlation should be included with the calculations.

Another technique is to perform a design inspection. This is usually accomplished by making a detailed examination of individual mathematical calculations, input values, correctness of assumptions, etc. Such a verification may be done for an entire set of calculations or by sampling different portions of the set. This approach focuses on confirming the accuracy of results, in addition to the adequacy of the approach. Obviously, this cannot be done in all cases due to the constraints of time, cost and availability of skilled reviewers.

One other approach is frequently used when the analyst's immediate supervisor serves as the verifier. The supervisor makes an overall review of the methods, assumptions and inputs used; asks several questions of the engineer about the process to verify that the engineer understood the proper use and limitations of the method and data; and checks the results in terms of order-of-magnitude, based upon the supervisor's experience with this type of product and its analysis. Although some critics contend the supervisor may not be knowledgeable, insufficiently independent to be objective, or just too

busy to do anything more than a cursory check, the author's experience is these concerns are not well founded. Supervisors typically
are knowledgeable and take the necessary time to make an acceptable
level of verification, especially for junior engineers or those engineers
who have some history of problems in the past.

When verification of a calculation is made, objective evidence of
the review should be provided for design assurance purposes. This
can simply be a notation and signoff on the document, such as: "Calculation verified (date and verifier's signature)." Another way is to
attach the sheet containing the simplified calculations to the original
calculation. The verifier should always sign and date his supporting
calculation sheets.

Another means is to develop and use a standard checklist for calculations. Figure 6.5 shows an example of such a checklist. The
verifier then fills in the necessary identification information for the
specific calculation, marks the appropriate items on the checklist indicating those items of the calculations which were checked, and signs
and dates the checklist. The checklist is then attached to and filed
with the calculation.

In those situations where a detailed inspection of a calculation is
made, it is a good practice for the verifier to make notations such as
"OK", "Acceptable", "I concur" or checkmarks in the margins adjacent
to the items or entries reviewed. In addition, the verifier must sign
and date the original calculation sheet as the verifier.

In a small company, with perhaps only one engineer, there are
still ways of providing some technical checks and balances. One approach is to employ the services of a local consulting engineer who has
knowledge and skill in the specific industry or with the general product
line. The consultant would then review the new design to verify the
adequacy and general correctness of the calculations.

Another approach is to make a cross-comparison of the calculations
with calculations performed on similar prior designs wherever possible.
This can provide reasonable confidence if the previously analyzed design has been used successfully in operation. However, the previous
design must be sufficiently similar to the new design to be valid for
comparison purposes. Evidence of such comparisons and correlations
should be included in the calculation records.

These, or similar actions, can be used to demonstrate some form of
verification of the calculation was made. It is an important aspect in
assuring the quality of design.

6.3.4 Technical Reports

Frequently it is necessary to document a design or analysis study in a
formal technical report. Such reports may be required by contract or
may be needed for distribution within the company as a technical ref-

DESIGN VERIFICATION CHECKLIST

Criteria	Verified (initials)
1. Requirements	
a. Customer special requirements incorporated in design.	M.J.
b. Conforms to company standard criteria	M.J.
c. Applicable codes/standards used in analysis.	M.J.
2. Methods	
a. Assumptions defined and considered reasonable.	M.J.
b. Appropriate methods used.	M.J.
c. Sample of input data checked for accuracy.	M.J.
d. Necessary iterations performed.	M.J.
e. Impact of variability considered.	M.J.
3. Results/Conclusions	
a. Results checked by	
1) alternate calculations, or	
2) comparison to prior designs, or	
3) design inspection of calculations	M.J.
b. Results within design limits	M.J.
c. Results support design requirements	M.J.

MICHAEL JAMES	5-6-83
Verified by	Date

FIGURE 6.5 Sample calculation checklist form.

erence. Since technical reports are often used for decision-making purposes by persons other than the report writer, the technical correctness must be assured.

The primary concern from a design assurance standpoint is the accuracy of the data and the validity of the conclusions. It is the quality of the information, not the manner of reporting, that is of real importance.

Thus, the engineering group needs to establish a process for review and approval of all formal technical reports prior to publication. Even though it is common to require some management signoff of such reports, many times the manager who signs is mainly concerned with the conclusions and recommendations, not with checking the technical accuracy or correctness of the data and calculations. For design assurance purposes, a technical verification review by one or more technically knowledgeable persons should be made as the report is developed. The purpose of such a technical review is the same as for calculation verification presented earlier in this chapter. The manager responsible for approving the report should verify that such a technical review was performed and should receive the results of the review before signoff. Any changes required as a result of the technical reviewer's comments should be incorporated prior to management approval of the final report.

6.4 DESIGN INTEGRATION

As products become more complex, the task of integrating the various components, subsystems and systems becomes increasingly important. This interaction of the items must be carefully designed and controlled for the system to operate properly. It is one more element of design that needs planned and systematic methods for implementation and control.

From the standpoint of design assurance, this aspect of the design process must not be overlooked. Several key elements from other sections of this book need to be applied to the design integration process. For example, there must be a clear assignment of engineering responsibility at each level of the component and systems design. The total set of requirements from the lowest to the highest levels must be defined, properly allocated and incorporated into the drawings and specifications. Specialized tools, such as interface control drawings, need to be developed and maintained. And as more and more levels of subassemblies and assemblies occur in the system, the change control process becomes increasingly difficult and important.

In fact, the entire process of engineering seems to be involved in the design integration activity. It is truly a test of the design assurance program.

6.5 SUMMARY

The design process requires an orderly approach, built on science and logic. Engineering departments need to use written procedures to define how the work is to be done and controlled administratively. In addition, the engineers should apply standard design methods when performing the technical design tasks to get the benefit of the proven technology for their designs.

A portion of the orderly process is the use of calculations to size components, verify interactions and investigate operational characteristics of the products. As such, these calculations really are the basis for the design and must be complete and accurate. The calculations need an independent review to minimize the chances of error or oversight, and the verified records must be retained for supporting the design.

The proper integration of a system requires many elements of a design assurance program to be used satisfactorily. It is indeed a challenging task. Nevertheless, it is one that can be accomplished through careful planning and development of basic design methods and attention to detail when implementing these methods.

7

Control of Engineering Software

7.1 Introduction   125

    7.1.1 Characteristics of Software   125
    7.1.2 The Software Life Cycle   126
    7.1.3 Controlling Software Quality   127

7.2 Software Design   128

    7.2.1 Definition of Requirements   128
    7.2.2 Top Level Design   129
    7.2.3 Detailed Design   132

7.3 Coding and Debugging Computer Programs   132

    7.3.1 Coding Practices   133
    7.3.2 Debugging the Software   136

7.4 Program Verification   138

    7.4.1 Integration Testing   139
    7.4.2 Acceptance Testing   140

7.5 Documentation and Control of Computer   141
    Programs
    7.5.1 Software Documentation   141
    7.5.2 Software Configuration Control   142
    7.5.3 Software Design Reviews   145
    7.5.4 Software Problem Reporting   146

7.6 Summary   149

7.1 INTRODUCTION

Computers are the way of life for most engineering departments today
and for virtually all tomorrow. The computer provides enormous capabilities to engineers for studying, developing, designing and analyzing products inexpensively. It offers new frontiers in innovation and
creativity. In fact, it seems to be limited only by the skill and imagination of its users.

However, the usefulness and effectiveness of the computer depends
greatly on the software which governs its operation. The software
provides the intelligence and control for obtaining the computational
and data-manipulating capabilities of these machines. Whether it is
one of the giant mainframe units, a mini- or personal computer, or a
tiny microprocessor, each one is directed and controlled by its applicable software.

Thus, the software package is the engineer's key for unlocking
the vast capabilities of the computer. Yet, it is this same software
which can be difficult to develop and exasperating to use. Now, let's
look at how software differs from hardware.

7.1.1 Characteristics of Software

Computer software, defined as those special types and forms of instructions which direct the operation of a computer, is different in
its nature and behavior than hardware with which engineers are very
familiar. Some of these differences are described below.

1. Hardware tends to degrade and wearout with age or use.
   Software doesn't. In fact, it usually gets better as latent
   errors are found and corrected.
2. Hardware generally gives advance notice of pending failure.
   When hardware fails, there is normally a clear sign or sound
   at the time of failure. Software doesn't. It can fail suddenly,
   without warning, and often invisibly. (Software failure is
   defined as an improper output or no output after receiving
   valid input.)
3. Hardware is made up of a finite number of parts and assemblies and have a relatively small number of failure modes and
   consequences. On the other hand, software typically has
   many, many paths which can, and frequently do, contain
   latent errors which can cause many different types of problems.
4. Hardware is visible and tangible, whereas software is often
   obscure and abstract.
5. The mathematical processes used in software can easily provide
   results (even though erroneously) which contradict the laws of
   science. However, hardware is destined to follow them.

6. There is considerable standardization in the design and construction of hardware items; yet very little standardization has been achieved in the design and construction of software.
7. Hardware can usually be restored to its original capability by maintenance and repair actions. However, the software maintenance and repair actions frequently change the original capability or cause a new set of problems.
8. And, finally, engineers and managers have developed a firm grasp of hardware concepts, but software still is largely veiled in mystery and intrigue.

7.1.2 The Software Life Cycle

In an attempt to remove some of the mystery surrounding computer software, let's examine a typical life cycle of a computer program. The design and development process of computer software has many parallels to a typical engineered hardware product. Figure 7.1 illustrates the major tasks in a software project.

As in any engineering project, requirements must be developed, identified and documented. Concepts are explored, and one or more preliminary designs are prepared. Some experimental work may be required to determine the feasibility of the concepts in meeting the design requirements. The more promising approaches are developed in further detail. Then, the design approach is narrowed, and the most promising design is selected for implementation. Design analyses and reviews are conducted, and finally the design is built and tested.

After it is demonstrated that the design meets the requirements, it is released for use. From this point on, the emphasis focuses on operation, maintenance and subsequent modification and upgrading.

The life cycle of a computer program can be very short, such as for simple, single-purpose programs (or those which do not work well, or are difficult or expensive to use, or are of questionable validity). Conversely, many versatile and well-designed computer programs can

DESIGN			CODING		VALIDATION		OPERATION	
REQM'TS	TOP LEVEL	DETAIL	CODE PREP.	DEBUGGING	INTEGRAT. TESTING	ACCEPT. TESTING	MAINT.	NEW APPLICA.

FIGURE 7.1 Software life cycle.

have useful lives of many years. Effective programs are commonly expanded and applied to new uses never conceived at the time the original software was developed. It is not uncommon to find that 50-75% of the total effort expended on a good computer program occurs in the operation and maintenance phase of its life cycle. Since this is a measure of quality and effectiveness of a computer program, let's examine some of the practices that contribute to longevity in computer software packages.

7.1.3 Controlling Software Quality

This chapter is especially written for engineers in a small to mid-sized engineering organization that are expected to develop and use their own computer programs. The methods and techniques contained herein are intended to aid the individual engineer in the design and development of software. It is not meant to be used by organizations with specialized staffs of systems analysts and computer science specialists, since these persons should typically know, understand and use these and other more sophisticated techniques. Nevertheless, the basic elements of this chapter apply to large computer programs as well as the smaller, day-to-day software projects.

One of the first and most important concepts of software quality to understand is that correctness and accuracy of the program must be built in. It can not be achieved by testing. There are too many paths and combinations in all but the simplest programs to be totally checked by tests. Therefore, the guiding principle is: "Do it right the first time."

This is best achieved by spending time and thought in the design stage and then progressing in an orderly and deliberate manner through the detailed design, coding, testing and integration steps. Many of the techniques which are considered as good engineering practices for hardware apply equally well to the design and development of software.

These practices include: careful definition of requirements, a methodical approach to design, documentation of the design as it progresses, the application of analyses and technical reviews to verify the correctness and adequacy of the design methods and concepts, the use of planned test programs and the identification and resolution of problems and deficiencies. As advocated in the other chapters of this book, the design and development of engineering software can benefit significantly from good planning and analysis and careful attention to detail in all implementation steps. The focus must be on defect prevention, because, in software, the task for finding and correcting defects is a long, tedious and often unfruitful one.

7.2 SOFTWARE DESIGN

The end quality of the software is heavily dependent upon the quality of its design. Time and effort spent in design is usually recovered several times over in the testing and application phases. If the design phase is neglected, many hours of debugging, testing and revising can be expected. This will be needed to unscramble errors caused by incomplete problem statements, inadequate design concepts, false or incomplete logic, weak program structure, inadequate program documentation, etc.

Although there are many specific methodologies described in the computer programming literature for proper software design, they tend to be grouped in three broad categories. These are: the definition of the software requirements, the top-level design and the detailed design. Each of these are explained in the subsequent sections.

7.2.1 Definition of Requirements

The starting point for any hardware or software design is to identify and understand the applicable requirements. As simple as it seems, this is often skipped over lightly.

Many of the methods and techniques described in Chapter 2 can also be applied to the software design process. Formal software quality standards, such IEEE Std. 730-1981, Military Specification MIL-S-52779 or MIL-STD 1679, require the preparation of a software specification. The purpose of having the specification is a good one; it requires thinking to prepare it and, when completed, provides objective evidence of the thinking process. Even for relatively simple programs or for small engineering departments, the planning and decision-making process of formulating a program specification is very helpful.

One approach for compiling the software design data is to use a three-ring binder or spiral-ring notebook to record and collect the decisions, requirements, assumptions and plans applicable to the development of a new software package. The notebook then serves as a working tool during the active life of the project. It provides the background for preparing formal documentation and serves as a reference for persons in the future who will be required to maintain or modify the finished computer program. Thus, the software notebook is an important record and should be prepared with care and preserved for later use.

Another technique for application during the requirements definition phase is the Technical Requirements Review. It is a form of design review which is conducted prior to starting the actual design of the software. The purpose of the review is to verify the adequacy and completeness of the requirements to be used when designing the software.

The review process can be formal or informal and may only involve two or three persons knowledgeable of the programming technology and of the needs for the particular program of interest. The engineer assigned the responsibility for developing the software can serve as both chairman and secretary of the review. The reviewers should have time to review the existing requirements and think about the needs of the software program. Then the reviewers should meet together with the software team leader and present their comments, concerns and recommendations. All such inputs should be recorded in the software notebook and space provided for noting the disposition of each (e.g., accepted and incorporated, deferred for further investigation, considered and rejected because . . . , etc.) The Technical Requirements Review only takes a few man-hours but can be an effective tool for preventing defects. It helps get the design phase started on the right track. Experience shows that having a comprehensive set of software requirements contributes significantly to good program design. This, in turn, minimizes false starts, oversights and errors in logic, and later in coding, due to inadequate understanding of what is needed.

7.2.2 Top Level Design

The main weapon for preventing software quality problems is an orderly, structured approach. Within the past decade, the concept of structured programming has evolved. It is widely accepted among the computer specialists as the proper way to design and construct computer programs. Although structured programming involves numerous technical actions, these can be grouped into a few broad categories.

The first of these is the concept of top-down design. This is directly comparable to the functional analysis method described in Chapter 2. The process begins by stating the purpose and objective of the software program. This serves as the top level of the program structure. Then the basic purpose is subdivided into the major functions that are required to be performed in order to satisfy the purpose statement. This first subdivision creates the second level in the structure.

The partitioning process is continued downward, one level at a time, until no further subdivision of the functions is needed to execute the program steps. The resulting structure then resembles a tree, where two or more functional blocks are subordinate to each higher-level block. This is known as a hierarchical structure and is the preferred output of a top-down design.

In contrast to this is the bottom-up design. This approach starts at the lowest detailed (machine interface) function and integrates horizontally and vertically upward. Although skilled programmers can, and do, use this approach, it tends to be inefficient. Also it makes it easy for latent defects to get into the program, as a result of faulty

logic or errors of omission. It also frequently causes the programmer
to patch over a problem. Rather than really fixing it in a logical,
straightforward manner, the programmer may be inclined to take
shortcuts or be "clever" in an effort to avoid scrapping a substantial
amount of prior programming which would otherwise have to be redone.

Bottom-up design also leads to the generation of network structures. A network structure differs from a hierarchical structure in
that no one module of the program is clearly superordinate to other
modules. The two conditions (tree and network) are illustrated in
Figure 7.2. The network structure tends to be intertwined in a complex manner which can be very difficult to check for correctness. It
is common to experience erroneous behavior in one part of a network
program after corrections or changes are made in another portion of
the program. The interactions of network structures can be very un-
predictable. Thus, the tree, or hierarchical, structure is much better
from the standpoint of software quality and ease of maintenance.

The process of subdividing or partioning of functions into smaller,
discrete tasks is the second major element of structured programming.
Each subdivision is referred to as a module and should be a well-de-
fined portion of the program. For best design, each module should
have the following characteristics:

1. it should carry out a particular program segment
2. it should be independent of other modules in performing its
   function, other than input/out
3. it should have a single point of entry and a single point of
   exit
4. it should have its own unique and descriptive name
5. it should be subordinate to only one higher level module,
6. its program statements should be limited to only those statements which are directly related to the function of that
   module

In the partitioning process, there are a few other rules that apply.
It is important to complete one level before beginning the partitioning
at the next lower level. By doing it in this disciplined manner, many
faults in logic or errors of omission can be avoided.

Another aspect of partitioning to consider is that of span of control. How few or many modules should be subordinate to any one
higher-level module? Most well-designed programs tend to have a
span of control of two to seven modules. This is not an absolute rule,
but experience shows that a very large span of control, greater than
about seven modules, is frequently a sign that a logic error was made
when constructing the higher level.

The third element of top-level design is to prepare a documented
description of the basic approach that is being pursued. This is
generally accomplished in a narrative text format, supplemented by the

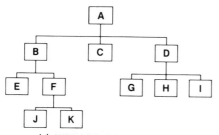

(a) HIERARCHICAL STRUCTURE

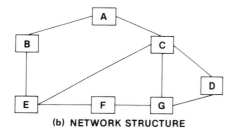

(b) NETWORK STRUCTURE

FIGURE 7.2 Types of program structures.

tree diagram of the basic modules and summary type of flow charts to show logic and data flow. The narrative should describe the purpose and program concept for each module shown on the hierarchical diagram. It should also define the various data inputs that are required, the output data, and the recommended formats for the output reports.

In addition to the design descriptions, a software test plan should also be initiated as part of the top-level design effort. The test plan should describe the approach to be followed when performing the various tests of the software. The tests should verify the software meets the design requirements as well as to flush out errors. This document will be finalized during the detailed design phase. For large programs, the Test Plan will be used in preparing detailed test procedures. For small programs it typically contains sufficient detail to be both the plan and the procedures. Regardless, the test plan is an important contributor to software quality.

This documentation then serves as the instructions for the detailed design phase. The output of the top level design provides the road map for moving from a broad problem statement to a series of smaller, manageable subtasks.

7.2.3 Detailed Design

The software designer's tasks of top-level design and detailed design
are analogous to the hardware designer's function of preparing layout
drawings (top-level design) and the subsequent generation of detail
and assembly drawings (detailed design). Thus, the detail designer
uses the top-level design documentation as the layout for the program.
The next task is that of developing the nitty-gritty details required
to make the program work.

Each module must be defined in specific terms of logic to be followed, data to be processed, decision steps to be accommodated, sequencing of operations, data storage, etc. This typically will include
one or more flow charts and a compilation of notes, definitions of
terms used, sketches of planned displays or reports and an explanation of special requirements or limitations that must be built into the
module. More experienced programmers often define the module requirements in Program Design Language (PDL). This is a specialized
technique that overcomes many of the problems caused by unclear
flow diagrams.

The output of the detailed design is then used by the programmers
to construct and write the actual code. However, a word of caution
is in order. There is a seemingly natural instinct for programmers to
want to run faster than they should. It is best described as the
"design a little, code a little" syndrome. However, this is contrary
to the concept of structured programming and is conducive to the creation of latent defects and inefficiencies in the program, due to inadequate thought and planning. Therefore, be faithful to the top-down
approach, using disciplined partitioning, level by level. Remember,
the best defense against program faults and defects is a systematic
and orderly design process. Do it right the first time.

If this philosophy is followed, so far no actual coding should have
taken place. However, the design is now well defined, and the coding
can now proceed in an orderly fashion. One or more programmers can
proceed with the coding, working largely independent of each other,
and accomplish the tasks in harmony with the overall program objective.

Let's now examine the coding activities.

7.3 CODING AND DEBUGGING COMPUTER PROGRAMS

Coding is both an art and a science. Over the years, much creativity
has gone into the development of programming languages. These
languages make it easy for people to give instructions to the computer
hardware. Yet, these programming languages follow specific rules
and conventions. Thus, each language imposes its own methodology
on its users.

Nevertheless, experienced programmers have learned and applied a series of techniques which minimize errors in the coding and documentation processes. Several of these tricks-of-the-trade are explained below. This description is not all-inclusive, but it does cover methods which can be followed, regardless of the size of the program, the language or machine used, or the end use of the program.

7.3.1 Coding Practices

The guiding principle in coding a program to minimize errors is to stick to basics. Several years ago, it was discovered that five fundamental constructs can address all code situations. These five constructs are illustrated in Figure 7.3. By applying these basic techniques, a

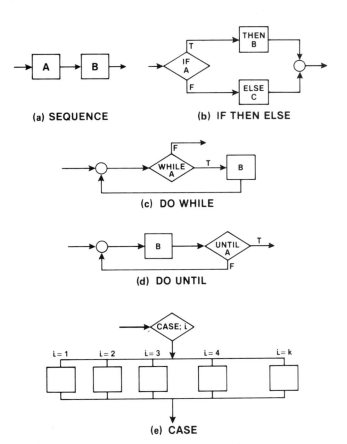

FIGURE 7.3 Preferred coding constructs. (From MIL-STD 1679, Weapon System Software Development, U.S. Dept. of Defense, Washington, D.C., 1978.)

programmer can build a program which is easy for others to understand and to maintain or modify in the future. Programmers should
resist the temptation of creating cute or clever (non-structured) solu-
tions for their immediate needs. Such non-standard methods nearly
always result in later problems in testing, subsequent correction of
defects, or in adapting the program to a new or expanded application.
Figure 7.4 shows the recommended nesting or sequencing of the
basic code constructs to achieve maximum accuracy and efficiency in
their application.

A negative rule which is a companion to the previous recommenda-
tion also needs to be emphasized in the coding phase. That is, avoid
using branching, or GO TO statements. This was a favorite technique
some years ago to reduce the coding effort, due to high machine costs
and limited memory size. However, experience shows its use is fraught
with horror stories, because the branching often creates unplanned
loops. These, in turn, result in program errors that frequently are
very difficult to find and fix.

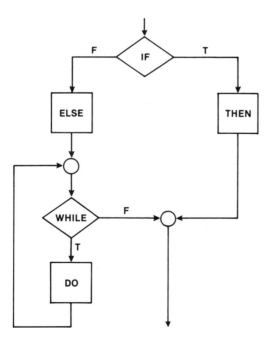

FIGURE 7.4 Preferred Nesting of Code Constructs. (From MIL-STD
1679, Weapon System Software Development, U.S. Dept. of Defense,
Washington, D.C., 1978.)

In the long run, even though using the five basic constructs may seem like more effort than using GO TO statements, the final program will have a higher level of quality and be easier to maintain or modify.

The next guiding principle is to select and use one of the commonly-understood high order languages. These include the well-known programming languages of ALGOL, BASIC, COBOL, PASCAL, and FORTRAN, as well as some of the newer languages, such as ADA, APL, CMS-2 and PL/1. These languages are well-known and understood, and their rules for use are clearly documented. Nevertheless, some programmers still like to write code, using assembler language. This language is not as well understood, requires more specific instructions, and its use raises the risk of creating program faults, both initially and in later operational use. In some cases, it may be necessary to use assembler language to interface systems which were previously written in different compiler languages e.g., COBOL and FORTRAN. However, programming in assembler or uncommon compiler languages should routinely be avoided for design assurance purposes.

Another important practice is to limit the sheer size of each program segment to a manageable amount of code. Typically, this means the code for any one module should not exceed approximately 50-100 executable statements or a maximum of roughly two handwritten pages of code. Some experienced programmers extend the limit as high as 200 statements but rarely would ever go beyond that.

These limits are intended to keep the module to a relatively small package to make it easier to understand and to demonstrate its accuracy and freedom from faults (surprise results). Both the original programmers and subsequent users can benefit from this.

There are several code-writing practices that are beneficial for making the code easy for others to understand, troubleshoot and adapt for other applications. The first of these is the generous use of comments statements throughout the coding. These statements should explain what is being done and why this particular logic is applied or action is taken.

Another recommended practice is the insertion of headings or titles for each of the modules and for labeling and identifying various subsets within the module code. Such labelling improves the understanding and traceability of the code and its logic. For the sake of consistency, use similar names for similar items.

Another technique experienced programmers have used is to format the code statements with a pattern of indentations of statements to show superior/subordinate relationships among code statements and subsets within the module. The printed code then resembles the format of an indentured parts list.

Thoughtful numbering of code statements is another technique which can enhance the understanding and traceability of the code. Instead of numbering the statements in a simple sequential fashion,

(e.g., 0001, 0002, 0003, 0004, etc.) choose number blocks which show
superior/subordinate relationships and allow for later modifications
without totally renumbering the code. For example, the first statement of the module might be 1000, the first series of subset statements,
1100, 1105, 1110, etc. The next series of subset statements might be
numbered 1200, 1205, 1210, etc. Combining this approach with the
indenting practices described above will provide a format that is easy
to read and shows a logical grouping of related statements.

As each module is coded, it should be subjected to a thorough
"desk-top" review by the programmer. Read the code statements
critically. Look for logic errors, program syntax mistakes, incomplete statements, etc., in the code. It is also a good practice to conduct a "design walk-through" of the code for particularly difficult or
unconventional modules. A design walk-through consists of an informal presentation of the code and its logic to one or more knowledgeable
programmers. They then comment on the acceptability of the approach,
validity of the logic, code construction, etc.

As the code is compiled, the compiler may identify syntax errors,
and the programmer can correct these. Sometimes, one statement or
portion of a module will be continually rejected by the compiler. When
this happens, it is a good practice to "dump" the module and get a
hardcopy printout for detailed examination. Compiling and debugging
is an interactive process which must be continued until all statements
in the module have been successfully compiled. A word of caution is
an order at this point. Successful compilation is only the first step
along the path to quality output. It does not mean the program code
is free from faults. Remember—neither the compiler nor the computer
hardware can recognize an error in logic. If the program says black
is white, the computer will accept it.

In recent years, several computer tools have been developed to
aid in the design, coding and testing of computer programs. Most of
these tools consist of specialized software packages which automatically check the accuracy and correctness of a program. Table 7.1
describes several of these. These tools are most generally needed and
used by software specialists on major program projects.

7.3.2 Debugging the Software

After each module is successfully compiled, the programmer is anxious
to see if the module will run and give the correct output. Obviously
one of the primary measures of quality in a computer program is that
it executes as expected and gives the proper output when valid input
is entered. However, the mere fact that it works once is no real
assurance that it will work with other combinations of data and operating conditions. Thus, the real task facing the programmer is that of
rigorously exercising the module to flush out the errors and built-in
faults in the module. And you can be sure there will be some.

TABLE 7.1 Tools for Preventing or Identifying Software Faults

---

File dump: provides listing of data in memory at time fault occurs

Change tracker: records each change to source code

Simulator: allows host computer to perform as a microprocessor would for testing embedded software

Program design language: A structured system for defining the design parameters in machine-recognizable language

Flow charters: provides diagrams directly from source code

Preprocessors: provides special checks of language prior to compiling and often provides messages for diagnosing causes of errors

Postprocessors: used to check specialized applications, such as input/output for computer-controlled machines

---

As mentioned earlier in the chapter, the software test plan defines how the program is to be tested. This includes the debugging stage for each module. The programmer must review the test plan and follow its lead.

One of the basic methods used in the debugging stage is to enter specially selected data, covering the range the program module is expected to process. This should include large, medium and small values to explore the range. It is often appropriate to include positive, negative and zero values, even if the program is intended to process only positive values. In addition, include some values which are representative of erroneous or troublesome conditions, such as an imaginary number or dividing by zero. It may also be appropriate to try subtracting the exact amount of an existing answer, which results in a zero remainder to see if the module will respond correctly.

Another typical debugging check is to input values with missing or incorrect units of measure, common fractions vs. decimal fractions vs. integers, numerical values in alpha fields, and non-numerical inputs for numerical fields.

The ideal debugging of a module would be to check every path and all combinations of data that the module will encounter. However, in all but the very elementary modules, consisting of only a few program statements, the number of paths and data combinations make this prohibitive. Nevertheless, it is desirable to exercise each of the paths in the module at least once during this checkout phase. This is to verify that the path will accept and execute when data is applied and that it does not contain obvious defects in logic or coding.

Debugging is an integral part of the coding process. It can be both aggravating and time consuming, but it is an important step in

TABLE 7.2 Common Problems in Coding

Missing or incorrect logic steps

Errors in use of language rules

Undefined variables

Typographical errors

Missing or incorrect commas, parentheses, or similar symbols

Failure to set or reset flags

Sensitivity to critical or limiting values

Exceeding data storage limits

Wrong addresses

Open-ended loops

Errors in use of programming standards

Numerical values in non-numerical fields, and vice-versa

---

the overall task. Table 7.2 lists many of the common problems which
occur during coding and must be debugged before the module can be
combined with other modules.

Upon successful completion of the debugging of each module, it
is then time to proceed with the integration of the modules into a working system.

7.4 PROGRAM VERIFICATION

The task of integrating the several or many modules into a working system is always a challenge. However, by following the top-down design
process with a top-down testing process, the task is considerably easier.

This is best accomplished in a phased manner. As described in the
previous section, each module must be tested to see if it executes correctly as an independent unit. (Module testing is often referred to as
unit testing.) Then, the task of integrating related modules into subsystems must be performed next. And finally the subsystems must be
integrated into a working system, including operation with the actual
hardware (e.g., computer, microprocessor, etc.) and other applicable
software. This approach builds on the orderly partitioning of the task,
but it requires planning and forethought to conduct testing in this manner.

The process of program verification is often performed in two distinct phases. These are integration testing and acceptance testing.

Integration testing is conducted as part of the software development process and is normally done by the analyst/programmer. Although it is a disciplined and rigorous process, it is relatively informal. On the other hand, acceptance testing typically involves participation by the end-use customer and is done in a formal and controlled manner.

For in-house programs, it is very desirable to involve the eventual user throughout the design and verification stages. This process can generally be more informal than if an outside customer is the ultimate user.

Each of the elements of program verification is described in subsequent sections.

7.4.1 Integration Testing

The guiding principle in integration testing is to proceed with deliberate caution. It should be performed in a top-down fashion to the maximum extent possible. Some lower-level integration of interfacting modules may be necessary, but routinely skipping over levels or pursuing a bottom-up test path is just asking for trouble and delays.

Top-down testing, applied one level at a time, verifies the top-level portion works before the detailed execution of all the modules is attempted. This is done by programming the modules as "stubs" (or black boxes). Then, when the interconnection between the superordinate and its subordinate modules is interrogated, the subordinate modules simply respond with a rote output to show the interconnections are correct. This is done instead of having the subordinate module initially execute actual data and serves as the first step in subsystem integration.

After the interconnections have been tested (and corrected as necessary), the process of subsystem integration proceeds in accordance with the test plan. Each module at the first subordinate level is then tested with a top level module to see if it will accept the preplanned test data, execute it properly and give a valid output. The process is continued as each module at that level is added and integrated with the other modules at that level, as illustrated in Figure 7.5.

The process is repeated for each successive level, first checking the interactions between superordinate and subordinate modules, and the integrating the subordinate modules, one at a time. As can be recognized, it is important to conduct the testing in a planned and disciplined manner.

The purpose of subsystem/system integration testing is similar to that of module testing: find and fix errors in the program. It serves a secondary purpose of seeing if the program will work. However, keep it clearly in mind that the main reason for this testing is to flush out errors in logic and coding. One of the ways of doing this is to use random inputs which are representative of "real-world" data. This approach will then exercise various paths and nodes and increase the

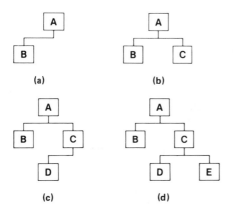

FIGURE 7.5 Integration of modules during testing.

chances of discovering hidden errors. Of course, this is in addition
to the standard tests for checking the program with large and small
values to explore the extreme conditions.

Systems testing should include running the full program on the
actual computer hardware it is intended to be used with in operational
service. Also the testing should be performed in the operational environment to the maximum extent possible. Only through extensive
testing can the inherent faults be found and corrected.

7.4.2 Acceptance Testing

For computer programs which are part of the package deliverable to
the customer, it is also necessary to plan and conduct an acceptance
test program. The purpose of this testing is to demonstrate to the
customer (either internal or external to the company) that the software
is compatible with the computer hardware, the software meets the specification requirements, and it produces correct results. This normally
must be done under the close scrutiny of the customer's representatives and performed in a controlled manner. Errors found in this phase
must be corrected and affected portions of the tests repeated to demonstrate the effectiveness of the corrective action. This process is continued until the rate of error discovery decreases below a pre-specified
level, or all elements of the acceptance Test Plan have been successfully
demonstrated.

A final word about testing—its main purpose is to find defects,
but experience shows that testing may not find all errors in a program.
Therefore, just because no errors are found in the final phase of testing, this does not assure that the program is now error-free.

7.5 DOCUMENTATION AND CONTROL OF COMPUTER PROGRAMS

The very nature of computer programs makes them abstract and, to a
large extent, intangible. However, for a computer program to serve a
useful purpose, it must be available in a form that can be understood,
controlled in a manner which keeps all the elements consistent, checked
to verify its accuracy and correctness, and monitored to understand
its reliability and relative freedom from errors. These are important,
but sizable tasks unto themselves. Let's examine some of the methods
which contribute toward assuring quality in the software.

7.5.1 Software Documentation

Several industry standards (i.e., ANSI/IEEE Std. 730, MIL-STD-1679,
MIL-S-52779) now define certain documents that should be provided in
the software package. Depending upon the application, the list can be
quite long. Government contracts are often very specific in defining
what must be delivered along with the actual program tapes.

For many industrial applications, the list can be generalized and
simplified as described below. These items provide the basic information needed to use, maintain and modify a typical computer program.

Requirements Specification: This is a compilation, in narrative
form, supplemented with tables and illustrations, of the program requirements. Format is not critical, but having a set of requirements
clearly defined is.

Software Design Package: The design package is the compendium
of design decisions and information used for constructing the computer
program. For large programs, or programs produced for an external
customer, the design package usually must be prepared and issued as
a formal report. For small programs, or programs of limited or only
moderate complexity, and to be used in-house, the software design
package can simply be the software designer's notebook. Table 7.3
lists the types of information that should be included in a typical software design package.

Source Code Listing: A listing of the compiled program code is
part of the standard program documentation package. As defined
earlier, titles, subheadings and explanatory comments should be used
generously throughout the source statements. For large or complex
programs, a program load map should also be provided to show the
location and size of each component of the program that are loaded concurrently in the computer memory or external storage.

Software Test Plan: The test plan defines how the program is to
be tested and explains the sequencing and rigor to be applied. It
typically includes the instructions for preparing the test procedures
and the selected data to be used in the tests. For large programs,
separate documents are required for defining the integration, acceptance and validation test processes. For small, or internally-used

TABLE 7.3 Typical Contents of a Software Design Package

Top-level design description (narrative)

Module hierarchy diagram

Design descriptions of each module

Data flow diagrams, flow charts or data processing tables

Definition of terms and data names

Estimates of memory requirements for various program
options

Sketches of displays and output reports

Notes, memos and meeting minutes, regarding program
design basis. (Should include decisions, assumptions and
limitations which affect use)

---

programs, it is still appropriate to prepare and use a written test plan.
This document should be prepared during the design phase.

  User Manual: The user manual must describe to a person who is
completely unfamiliar with this program, but is trained in working with
computer programs, how to use it effectively. The user manual needs
to describe the general approach, capabilities, limits and special requirements of the program. The user must be instructed what inputs
are needed, where and how to enter them and what outputs and options are available for use. It should also include information for
troubleshooting some of the more common causes of an aborted load run.

  The software documentation package is crucial in the actual control of any computer program. As such, it is extremely important to
apply rigorous document control practices for identifying each version
and revision of the hardcopy documents and source code tapes. Further, it is imperative to update the supporting documents promptly to
reflect changes in the code and to keep all hardcopy (documents) and
softcopy (code) consistent. Otherwise, design control will deteriorate
rapidly and irretrievably.

7.5.2 Software Configuration Control

For design assurance purposes, it is necessary to establish and maintain
tight and accurate control of computer programs. The task is even
more difficult for software than it is for hardware, due to the low visibility of code and the potentially-volatile impact of changes. Nevertheless, many of the methods and techniques described in Chapter 5 also
apply to software.

The basic tools are the use of baselines, formal control of changes, and a process of status accounting of approved (or pending) changes and their implementation. However, there is one important difference in the configuration control of software versus hardware. In hardware projects, configuration control practices are normally implemented late in the development cycle or at the beginning of the production run. In contrast, it is necessary to apply configuration control methods very early in the software development cycle. Otherwise, it will become impossible to define the baselines for control purposes. For software, configuration control is a powerful defect-prevention tool.

The software documentation package described in Section 7.5.1 constitutes the software baseline as it evolves. Each of the elements of the software package must be clearly and uniquely identified as it is produced and modified. This means each revision, regardless of how small or large the change, must be identified. For hardcopy documents, such as specifications, design reports, flow diagrams, this is usually accomplished by assigning a document number and a new revision letter for each change. Although the revision status of these documents can be verified by inspection, it may be tedious and time-consuming. However, for the softcopy (code), it is much more difficult to check each source code statement, compiler configuration, object code tape, etc., after the fact, to determine its configuration. Thus, rigid and disciplined practices need to be applied from the beginning to record each change. This typically is done by logging in and compiling changes as comments in the program code and by maintaining a master index for the document package, showing the unique and traceable identification markings for each item.

As the development of the program moves along, it is also necessary to implement a controlled process for defining, evaluating and approving planned changes. The effects of the change must be determined in advance of implementing it. This also allows setting the necessary action in motion to update the supporting program documentation to maintain consistency in the software package. Figure 7.6 illustrates a change status log which can be used for this purpose.

In addition, a specialized method, called Library Control, is required for nearly all programs. Library Control consists of giving the master copy of the software, including the latest official version of the code tapes, to one person or group. This is usually the computer operations organization. The librarian then serves as the custodian of the package for operation and change control purposes. Only the library version can be used for production runs. The librarian should routinely be involved in the implementation of all subsequent changes. In addition, the librarian is often responsible for, or is a contributor to, the configuration status accounting system.

During the testing of a new computer program, a large number of changes typically will have to be made to correct errors flushed out by the various tests. This is where a good configuration control system

Date: 1/31/84

CHG. NO.	DESCRIPTION OF CHANGE	REASON FOR CHANGE					APPVD		IMPLEMEN.		STATUS	
		ERROR	CUST.	HDW.	MOD.	OTHER	YES	NO	CODE	DOC.	TEST	OPER.
381	ADD MODULE FLTLO—AIRCON PROGRAM		X				X		X	3/16	4/20	6/1
382	REVISE TEST PLAN FOR CUST. ACCEPT. —CHARTS PROGRAM	X					X		N/A	3/1	3/31	N/A
383	REVISE MODULES SPINS & HEATR TO ACCOMM. MOD. 2 TO CITS COMPUTER CHIP			X	X			X	4/6	5/18	6/20	7/31

FIGURE 7.6 Software change log.

pays for itself many times over.

In the testing phase, the programmers will often use patches to correct an error condition to keep the testing moving forward. Patches are quick fixes and may not be written in final form in the program listing. As easily imagined, temporary patches can be a source of errors and inconsistencies. Therefore, all corrections must eventually be properly integrated into the software documentation package. Otherwise, it will be impossible to know or prove what program configuration was actually tested. This could have a serious impact on the final acceptance and validity of the computer program.

Configuration control of computer software is another example of where administrative and clerical tasks become entwined with the scientific and technical aspects of engineering and product design. The challenge is to develop and use effective administrative practices which aid, not impede, the design process.

7.5.3 Software Design Reviews

Another means of enhancing software quality is through the use of specialized design reviews. In Chapter 9, the topic of design reviews is discussed in considerable detail. Many of those techniques can also be applied to software design efforts.

Several different types of design reviews can logically be used at various points in the software development cycle. Each of the different types are described below.

Desk-top Inspection: A desk-top inspection is normally performed by the programmer. It is a step-by-step review of program design features, program logic and/or source code statements. The purpose is to make a comprehensive review in search of errors in logic, program syntax, omissions, etc., before committing the software to the next phase of development. Desk-top inspections should be routinely performed on each module prior to compiling new code statements and before each module test.

Code Inspections: A code inspection is similar to a desk-top inspection, except it is normally performed by one or two other experienced programmers. They first review the requirements specifications and software design definition documents. Then they examine the source coding to judge whether it is in compliance with the software requirements and design concepts, if the approach is reasonable, and if there are any noticeable problems or concerns about the logic or coding. The review is generally informal with the results fed back directly to the responsible programmer. Even though the reviews may be informal, it is a good practice to make notes of the review comments and their resolution in the software notebook.

In major programs, such as under government contract, specialists from the contractor's software quality assurance group and government representatives may participate in code inspections. When this occurs, the process is normally more formal and the results of the review are documented in the design review minutes.

Formal Software Design Reviews: Formal design reviews can be very helpful for assuring quality in the software design activities. Formal reviews are often conducted at the following stages of program development:

1. Review the requirements specifications prior to performing the software design.
2. Review the top-level design documents before proceeding with detailed software design.
3. Review the final design package before performing the program coding.
4. Review the program management plans before the end of the design cycle (e.g., configuration management plan, integration test plan, acceptance test plan).
5. Review the program documentation and status prior to release of the computer program for operational use.

The details and methodology for planning, staffing, conducting and closing out formal design reviews contained in Chapter 9 can be adapted with only minor changes for the software effort.

Informal Design Reviews: An informal review can be used in lieu of, or in addition to, the formal design reviews described above. The primary differences between the formal and informal reviews are: typically fewer people participate in the informal reviews; there isn't the structured meeting led by an appointed chairman; and meeting minutes are not prepared and distributed. Nevertheless, for small projects or small companies, informal design reviews can be an effective tool for avoiding errors, oversights or misunderstandings during the software design and development effort. Typically, two or three heads are better than one. Again, it is a recommended practice to record the significant comments and their resolution in the software notebook.

7.5.4 Software Problem Reporting

Another tool for measuring and evaluating the quality and reliability of a computer program package is the Software Problem Reporting System. As stated many times before, new software is fraught with errors that occur for many reasons and come from many sources. Although the primary thrust is to prevent errors and program defects by doing it right initially, latent errors do slip in, and subsequent corrections, changes, and modifications can introduce more problems.

The vast majority of errors are eliminated through the compiling and debugging stages. However, some errors usually remain when the program is released for integration and acceptance testing. Even after final acceptance of the program, there is a high probability that some errors and faults still are contained in the software that may later cause problems or unpredicted behavior in operational use. The software problem reporting system provides a vehicle for capturing and evaluating information about program faults and failures.

The software problem reporting system is similar in many respects to the approach used for controlling non-conformances in hardware. (Ref. Chapter 11). But there is one significant difference. Software defect trends are frequently used in the acceptance process to judge whether or not the software is sufficiently free of problems to be considered acceptable for operational use. Figure 7.7 shows representative patterns of failure trends that occur in practice. Figure 7.7a illustrates what often occurs in a well designed and structured

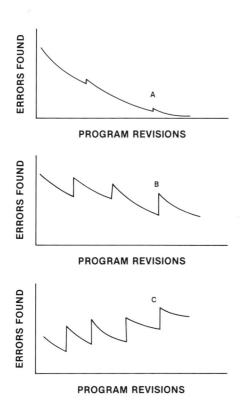

FIGURE 7.7 Software fault trends.

program, i.e., the number of defects found are decreasing with time. In Figure 7.7b the pattern represents a complex program which is still filled with logic and coding problems. Note that the frequency of defects discovered remains relatively constant. Figure 7.7c is an illustration of what happens when efforts are made to modify a program which was originally designed and documented poorly, i.e., new problems are being created in the validation or modification phases faster than they are being removed.

The major elements of a software problem reporting system are the disciplined reporting of software problems on a predetermined report form and the compilation and analysis of the reported problems. The resulting data is then used as a measure of program quality and reliability and, of course, for current and future corrective action.

The exact content and format of the problem report is not of great concern; however, it needs to contain enough information to define the type of problem, when and where it occurred, the impact of the problem, etc.

The problem data should be analyzed periodically to decide what, if any, corrective action should be taken, and where it is needed the most. For instance, monitoring the total number of problems found in one program doesn't tell much, but a tabulation of problems by module could be very revealing. Also a breakdown of all problems found on all programs in a given period of time by software life cycle phases (requirements, top level design, detailed design, coding/debugging, integration testing, etc.) could identify where additional supervision, training or tools might be needed.

In some situations, the rate at which errors are found and corrected during acceptence testing may provide a measurable means for judging when testing can stop. The error rate may be established in terms of mean time between errors, or X errors per 1000 lines of executable code statements or similar numerical limits.

Software problem reporting, however, faces a difficult paradox, especially during the development phases. The people causing the problems (the programmers) are the ones that find the problems they created and must report on themselves. Therefore, effective implementation of software problem reporting requires a considerable amount of tact and discretion by management. Otherwise, feelings of self-incrimination on the part of the programmers will reduce the effectiveness of the reporting system.

In the long run, the reporting and analysis of problems can be a constructive process and lead to improved quality in the software. It is well worth the effort.

7.6 SUMMARY

More and more engineering work is being done on the computer. And more and more engineers are doing their own programming. But computer programs are less visible and more abstract than the old familiar tools of engineering. Thus, it is much harder to know if the programs are right.

New mathematical modelling techniques are becoming very complex. In fact some observers have suggested that very large computer programs may be the most complex structures devised by man. In any event, the complexity of computer software makes it prone to faults and errors. Its sheer size and complexity frequently makes it impossible to verify the accuracy and correctness of each mode, path, and data combination.

As a result, the only hope of assuring quality in computer software is to concentrate on preventing defects from getting into the program from the start. By applying rigorous discipline, especially in the software design phase, and using proven techniques, there is a high probability of developing programs with the necessary accuracy and effectiveness. Some of the important techniques to use are: top-down design, structured programming, high order languages, independent partitioning, design reviews, top down testing, comprehensive program documentation and configuration control. These methods, coupled with new and expanding programming tools, can avoid the software problems that occur from an unstructured approach.

Computer analysis offers great opportunities to every engineering department. But it also poses a challenge to the engineers and their managers to develop and use effective software in their continuing pursuit of excellence.

8

Product Testing

8.1 Introduction   151

    8.1.1 Types of Verification Tests   151
    8.1.2 An Integrated Approach   153
    8.1.3 Test Cycle   154

8.2 Initiating the Test   155

    8.2.1 Test Requests   155
    8.2.2 Test Planning   156
    8.2.3 Test Procedures   157

8.3 Performing the Test   159

    8.3.1 Training of Personnel   159
    8.3.2 Calibration of Instruments   159
    8.3.3 Data Collection   159
    8.3.4 Running the Test   160
    8.3.5 Control of Nonconformances   161

8.4 Reporting Results   161

    8.4.1 Test Report Preparation   161
    8.4.2 Review and Approval of Test Reports   163
    8.4.3 Test Records   163

8.5 Summary   165

8.1 INTRODUCTION

Testing has long been used for demonstrating the adequacy of a product. Because testing gives direct, tangible results, it is an important means of showing what a product or design can do. However, testing is often expensive, and there are practical limits to the degree of realism that can be achieved when testing large or complex items. Nevertheless, testing is a basic method for verifying design adequacy.

Many different types of tests are used in industry today. Yet, nearly all tests have a common purpose, i.e., to obtain physical information about the product or system. This is especially important as input to the design process. Tests provide answers to questions, such as: Is material A better than material B in this application? Will design X withstand the required load? Is system Y capable of operating for the specified time in this environment?

Tests are used to investigate ideas, check the feasibility of a concept, or demonstrate compliance with a specific requirement. In each situation, the accuracy and validity of the test results have an impact on the quality of the conclusions. In fact, incorrect test data can be worse than no results, since erroneous data can lead to incorrect decisions.

8.1.1 Types of Verification Tests

Among the more common types of tests used by engineering departments are scale model tests, development tests, prototype tests, proof tests and acceptance tests. Each of these have a special role to play and are briefly described below.

Scale-model tests are frequently performed to evaluate new concepts. The tests typically are conducted on reduced-size components to keep costs down. Scale models are an effective means of examining new ideas where the primary purpose is to gain insights and determine feasibility. Some caution must be exercised, though, since the appropriate scaling laws must be selected and properly applied to the data before firm conclusions can be reached. Due to the preliminary nature of designs tested via scale models, such tests are only occasionally used for design assurance purposes. Nevertheless, scale model tests will continue to play an important role in the engineering process.

Development tests are a broad, generic category of tests which are used to explore the variables of a new material, component or system and refine it to a practical product. Development tests cover a wide range of methods and approaches. But all tend to follow the iterative process, i.e., learning from failures and building on successes. Such development tests are frequently used for design verification purposes. It is part of the engineering feedback process which influences the growth and maturing of a design or a product.

On the other hand, prototype tests are normally performed on full-size, preproduction items. These tests investigate the workability of the design prior to committing major resources for production. Items to be tested are usually made in a model shop, using temporary tooling and considerable hand crafting. Prototypes are close resemblances of the final design and represent the output from the design and development efforts. The test articles are normally loaded to the design conditions, and occasionally beyond, to demonstrate the design capability and to verify the correctness of the design and analysis methods. Such tests are important for verifying design adequacy.

Proof tests are usually conducted on items which are advanced prototypes (second or third generation designs or initial production units.) Extensive tests are then performed to demonstrate the limits of the product. Loads and operating conditions are usually increased progressively during the test until failure is reached. The limiting (failure) conditions are compared to the design operating conditions to determine if there is adequate margin in the design for the planned application. Several of the items are normally tested to examine the degree of variability present in the product. Results from multiple tests also aid in identifying the failure modes and weak links in the design. Proof tests are an important source of data for design verification.

Acceptance tests are a special category of tests which are used to demonstrate that the product meets the design requirements within its expected operating envelope. Acceptance tests are nondestructive by intent and are conducted on the final production product. Items which successfully pass the prescribed acceptance tests are usually delivered to the customer. Large or especially—complex or costly items are often subject to acceptance tests which are witnessed by a customer representative as a condition for final acceptance. This type of test obviously is a direct effort to verify the design.

In some situations, a special type of acceptance test is used. It is known as a Model Test, such as done in the aerospace industry. Subsequent production depends upon successful completion of the Model Test.

In a Model Test, one or a specified number of test articles are subjected to a rigorous series of tests to demonstrate the adequacy of the design. Often, the test requirements are defined in the customer's contract and frequently include a post-test teardown of the item to inspect for breakage, wear, deterioration, etc. Customer representatives typically witness the testing and participate in the inspection of the items after testing. It is common practice for the customer to withhold approval, and possibly funding, for production, pending the successful outcome of a Model Test.

8.1.2 An Integrated Approach

A comprehensive test program is an important building block in demonstrating the adequacy of a new design. This section illustrates how the
various types of tests can be integrated into an effective design verification program.

During the development phase, materials, parts and assemblies which
which are critical to successful operation or items which have a high
degree of risk or uncertainty should be identified and investigated by
individual development tests. The development tests should be aimed at
checking the validity of the basic design methods and confirming the
design assumptions. Such tests should be carried through until failure
occurs.

As the design progresses, prototype or initial production parts and
assemblies should be tested under simulated operating conditions to explore component interactions and the combined effects of environment
and loads. Also these tests must investigate the adequacy of the
changes made as the result of problems discovered in the development
tests or in the evolution of the design.

It is also appropriate to test the complete assembly (end item furnished to the customer) as part of the design verification process.
Wherever it is economically feasible on a cost-risk basis, the complete
assembly should be subjected to primary loads and environmental conditions up to specified operating values. If there are important secondary loading conditions that are significant to the safe and reliable operation of the product, these should also be checked, in addition to, or
in conjunction with, the primary loads.

Such tests are especially useful for design verification if they eventually can be carried to failure, although it may not be economically
possible in all situations. Nevertheless, testing to failure is valuable
for checking design predictions and identifying the weak link (part or
assembly most likely to fail first) in the product.

Completed units should be subjected to acceptance tests prior to
delivery to the customer. For items which are produced singularly or
in very small quantities (such as complex machines, or very expensive
apparatus), it may be necessary to conduct an acceptance test on each
unit to demonstrate compliance with key design requirements. In most
cases one or a small number of units can be "type" tested, i.e., tests
performed to demonstrate the generic design adequacy. Subsequent
production units then may only have to pass certain tests, or the production lots may be tested on a sampling basis.

Finally, the type of product or its application may dictate the need
for reliability testing to demonstrate life performance characteristics.
This form of testing can be very expensive and require a lot of time to
complete. However, it does provide data that is valuable for design
assurance purposes.

One final comment—do not underestimate the usefulness of full-scale testing of the complete unit or final assembly. Many important characteristics can only be evaluated in the final assembly condition, although tests of this type may be more difficult or costly to perform. Nevertheless, testing of the complete assembly is generally needed for design assurance purposes. It should be included in the design verification test program to the maximum extent possible. More on this type of testing is discussed in Chapter 14, Reliability Improvement.

The remainder of this chapter describes the recommended processes and practices for planning, conducting and reporting the various types of tests.

8.1.3 Test Cycle

Regardless of the type, nearly all tests follow a similar pattern from initiation through completion. Figure 8.1 illustrates the important steps that are typically involved in the test cycle.

For parts and materials, the cycle may be very short and simple. It may be possible to initiate and complete the tests in a matter of hours. On the other hand, a complex, high-technology system, such as a new spacecraft, may require months or even years to plan and perform the necessary tests. However, the major elements of testing are the same.

Each of these key ingredients are described in the sections to follow.

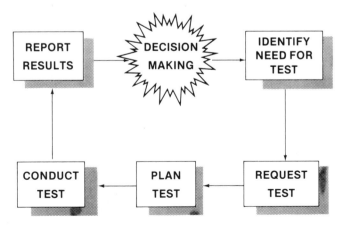

FIGURE 8.1 Typical test cycle.

8.2 INITIATING THE TEST

The design engineer is usually the person that identifies the need for
a test. The purpose may be to obtain data to be used in the design
process or to prove that the design will work in the real world. Regardless of the intent, the test requirements must be defined clearly in
some form of test request.

8.2.1 Test Requests

Although requirements for testing can be given verbally, a written test
request should be used. Experience shows that most tests are just too
complicated to be planned and controlled by verbal instructions. Table
8.1 lists the type of information that should be included in a test request. This can be tailored to specific needs as required by the complexity of the product or of the data needed.

For conceptual tests, the test organization may have to be consult-
ed to determine what type of test or test object can best provide the
requested data in a timely and cost-effective manner. However, where

TABLE 8.1 Typical Contents of an Engineering Test Request

Purpose of the test

Scope of test and intended use of the results

A description of the item to be tested (by sketch, part
number, etc.)

The operating and environmental conditions which must be applied or simulated (loads, temperature, pressure, voltage, etc.)

Number of tests or cycles to be run or number of items to be
tested.

Length of time or duty cycle

Specific data requested

Desired accuracy of results

Number of data points to be gathered at special or critical
conditions

Acceptance criteria (where known in advance)

Definition of any special conditions, limitations or critical
aspects to be considered in the tests

From: J. A. Burgess, Quality Assurance in Testing, Quality, Jan.
1983, Hitchcock, Wheaton, IL.

the requester needs a specific part or product tested, the requester must define the item to be tested by part number, model number, etc.

Since testing can be expensive and time consuming to set up and run, the test requester should indicate the approximate number of tests needed. Otherwise, the testing group may do too many or too few.

It is also a good practice to require each test request to be reviewed and approved by at least one higher level of management before issuing it to the test organization. The test request should be checked for need, clarity and completeness. The test request should then be dated, signed off and issued. Copies should be retained in the design files for later reference and use.

8.2.2 Test Planning

After receipt of the test request, the test group should review the request and develop a plan for obtaining the desired information. The test planning includes determining the best methods for running the test, defining the methods of measurement, and selecting the facilities and equipment to be used. In addition, the test group should identify areas of risk or uncertainty.

After this work is completed, the key test personnel assigned to this test should meet with the test requester. They should jointly review the planned approach and resolve any questions or differences. This is also an appropriate time to review the selection of facilities and the scheduling of the test.

All parties involved must recognize that the adequacy and accuracy of test results are strongly influenced by the equipment and facilities available to perform the tests. It is not unusual to find there are major limitations in what tests can be performed, conditions simulated or measurements made with existing equipment and facilities. Since new test equipment and facilities typically are expensive and have long procurement lead times, it may not be possible to do the necessary tests at company facilities. Thus, a decision may have to be made early in the planning phase on whether the tests can be done in-house, the task subcontracted, or the test request rewritten to match the available capabilities.

If the decision is made to subcontract the test, it is generally preferable for the in-house test group to handle the subcontracting effort, because of their familiarity with the details of testing. Preparation of the test specification, identification of testing sources, and the issuing of requests for quotations through the purchasing channels should be accomplished promptly by the test group, with periodic consultation and review with the test requester. Having tests performed by outside agencies is a common practice today, but it does add some rigidity and complexity to the process, e.g., contractual agreements, geographical separation, and formal communications channels, to name a few.

After the test group and test requester reach an agreement on the test approach, facilities to be used, interpretation of unclear areas in the test request, etc., it is a good practice to prepare a written summary of the results and conclusions from the review. Both the requester and test leader should jointly sign the meeting minutes, since this is an important document for design assurance purposes. This is especially true if it contains new or different information than contained in the test request. A copy of the minutes should become part of the test records file.

After the planning review meeting, it is then the test group's responsibility to proceed with the detailed implementation of the test. For purposes of clarity, the remainder of this section is directed at those steps which are necessary for accomplishing the tests, using in-house facilities. Nevertheless, the same requirements apply to sub-contracted testing, but the division of labor among the in-house test group and the subcontractor must be worked out on a case-by-case basis. In addition, all such arrangements need to be defined in the subcontract document.

Detailed design of the test includes the design and/or procurement of the test articles (items to be tested), the design and procurement or manufacture of the test fixtures for mounting the test articles, the design or selection of facilities and equipment to apply the required loads and simulate or reproduce the specified environmental conditions, the selection or procurement of instruments, gages and recording equipment to make the required measurements, and the selection/application of necessary safety devices to protect the testers and the equipment. All of these actions must be accomplished before any actual testing can begin.

8.2.3 Test Procedures

In addition to obtaining the necessary hardware, the test group must prepare and issue the test procedures which define how the test will be conducted. The need for test procedures varies with the type and scope of the engineering tests to be performed. The greater the complexity of the tests, the greater is the need for test procedures. However, don't overlook the benefits of using written procedures for even simple tests. Procedures represent thinking and planning. As a result, using written test procedures generally enhances the quality and accuracy of the test output. In addition, it frequently improves test productivity, e.g., optimum sequencing, shorter test setup times, proper selection and use of test equipment, and better data recording practices. Table 8.2 lists the basic ingredients of an effective test procedure. Obviously, the exact format can, and should be, adapted

TABLE 8.2 Basic Ingredients of Test Procedures

Clear statement of purpose and scope of tests (what is to be done and why).

Definition of items to be tested (in specific terms of part number, model number, material type, size and shape, etc.).

Precautions to be observed (identify hazards, safety considerations, special factors, such as degree of cleanliness of test containers, etc.).

Special Conditions/Limitations important to the tests (e.g., "record the impact test with high-speed movie camera"; or, "test requester must be present when voltage test is performed.").

Define data recording requirements (specify data sheets to be used, frequency of readings, data channels, etc.).

Specify how test is to be performed (be thorough in defining the application of loads, environmental conditions, sequencing of steps, and number of times data is to be taken at a particular step.).

Identify what actions are to be taken in the event an anomaly occurs (what to do if something unexpected happens, e.g., premature failure).

State what to do after completion of the test (disposition of test articles, teardown of setup, etc.).

---

to meet the local needs, but the items listed in the table should be considered when preparing a specific procedure.

Persons familiar with the testing process and knowledgeable of the equipment capabilities should be assigned the task of writing the necessary procedures. All elements of the requirements contained in the test request, as modified by the test planning meeting minutes, must be addressed.

It is also important for a procedure to progress step-by-step in a logical manner to accomplish the test objectives. It should define what action is to be taken, what is to be measured, how it is to be measured and when, how many data points to be taken, and how the data is to be compiled.

After preparation, the test procedures should be reviewed by at least one other knowledgeable person in the test group. This provides a check on correctness of approach, thoroughness of coverage, and adequacy of the test equipment and facilities for achieving the desired results. Whenever possible, this should be someone in the test group's management.

Some organizations go as far as to have the person requesting the test review the procedures prior to final issue. This provides one more check point along the path for obtaining valid results for design verification.

8.3 PERFORMING THE TEST

Up to this point, the emphasis has been on the planning of the work. Now it is time to work the plan.

8.3.1 Training of Personnel

It is the responsibility of the test group to provide whatever training of the testers is required. Training should include: operating the equipment, making various types of measurements, recording the required data, and judging whether or not the acceptance criteria is satisfied. Also, the test personnel must be trained in safe practices and know what to do in the event unusual or emergency conditions are encountered during testing.

Some test groups find it beneficial to conduct continuing, on-going training/retraining programs for their test personnel. Experience shows that time and effort invested in training typically pays off in improved practices and results.

8.3.2 Calibration of Instruments

The validity of test measurements is another area of great interest. The calibration status of measurement and test equipment should be checked closely. It is the responsibility of the test group to verify that the instruments and gages are properly controlled and calibrated. The use of measurement equipment that is out-of-calibration, or even of uncertain calibration status, must be avoided. There are many separate sources of information in the quality assurance literature, defining methods and practices for assuring the proper control of equipment calibration. As such, the topic will not be covered in this text. Nevertheless, the importance of properly-calibrated measuring equipment must be recognized. Persons using results from engineering tests for design assurance purposes have every right to assume the data was generated with properly-calibrated equipment, and the results are correct within the stated accuracies.

8.3.3 Data Collection

Data collection is another important step in the testing process. Again, test planning efforts should cover the process of what data is to be collected and how to record it. Although more and more automatic data acquisition equipment is becoming available, the majority of tests still require the manual recording of test results. Including specific data forms in the test procedures goes a long way in getting the testers to obtain the necessary data and recording it in a manner that makes it easy to use.

One other aspect of data recording should be noted. Test records must be clear and complete. They must provide a traceable path between the test data and the items tested. For example, the test records should include:

1. Specific identification of the items tested
2. The date the test was conducted (include the time of day if necessary)
3. The names or initials of the persons performing the tests
4. The particular test facilities and equipment used, (type of equipment and manufacturer's name and model number as applicable. List serial number if more than one is used locally, etc.)
5. Comments describing test conditions, unusual circumstances, etc.

Clear traceability is crucial to support later evaluations of proposed design changes and to support the design in the event future problems or questions about design adequacy should arise.

Finally, the matter of variability needs to be considered in the planning and performance of design verification tests. Granted, one test may be worth a thousand opinions. However, one test or one data point may also be misleading. It may be a maximum or minimum value— not really representative of the typical or nominal condition. Some repetition of tests is needed to examine the inherent variability of the results. Several basic statistical tools presented in Chapter 10 of this book can be used for assessing this variability.

8.3.4 Running the Test

Although it sounds obvious, the tests should be conducted in accordance with the test procedures. Unfortunately, this is not always done. The testers should be given an orientation, covering the purpose of the test, methods to be followed, and any special items they should be aware of. Simply giving the test procedure to the tester and telling them to "go to it" is seldom adequate.

The test procedures provide a pre-planned sequence and order for gathering data, and the testers should not deviate from these methods arbitrarily. However, the use of procedures does not prevent intelligent modification of the tests when new circumstances arise. If unexpected problems are encountered or the results turn out to be a lot different than expected, changes should be made. But such decisions

to change must be made by the test leader or test requester, not by
the testers themselves.

Since the early test results often are indications of things to come,
it is a recommended practice for the test leader and/or test requester
to be present during the initial or crucial phases of the test cycle.
They can provide on-the-spot answers to the testers' questions, see
the data and the test article performance as it is occurring, and recognize when there is a need to change, or at least to re-evaluate, the
planned approach.

When, during the course of testing, it becomes apparent that a
change in the established procedures is required, such changes should
be specified and authorized in writing by the test requester. All such
"mid-course corrections" should be included in the test files.

8.3.5 Control of Nonconformances

During the conduct of a test, it is not unusual to encounter an anomaly
in the data or an unexpected failure. It may be a design problem, an
error in the test method, a limitation of the test facility or equipment,
a malfunction of the equipment, or an external event, such as loss of
commercial power. Sometimes the cause of the problem may be obvious,
sometimes it may be quite elusive. Nevertheless, it must not be ignored. Efforts should be made to identify the cause of the unexpected
event. Even though it may be difficult, or impossible, to isolate the
true cause, a record of the investigation should be made, and a brief
discussion of the event should be included in the final test report.

If the tests reveal problems with the material or product, such as
premature failure or improper operation, this information must be made
available to the designer. Corrective action, such as redesign or
change of application, must then be taken. It is simply sound engineering to follow up on all open items or shortcomings discovered during
the testing process. This is the specific responsibility of the design
group.

8.4 REPORTING RESULTS

The final output of the testing effort is some form of test report. It is
the responsibility of the testing group to document the test process and
the results. Preparation, review and approval of the test report is
covered in this section. Also, the compilation and retention of test
records is discussed.

8.4.1 Test Report Preparation

A test report may be as simple as the completed test data sheets or as
comprehensive as a formal, bound document with narrative text, tables

of data, and illustrations of the test set up. For maximum effectiveness, the test group should prepare a report which presents the data in easy-to-understand fashion. Various texts on technical writing describe several different report formats that can be used. However, the inverted pyramid format listed in Table 8.3 is the one preferred in industry today. It summarizes the results and conclusions at the front of the report for management overview and provides the detailed description of the test, the test data, and the discussion of the results in later sections of the report for the technical and professional readers. However, any format that presents the results in a clear manner should be acceptable for design assurance purposes.

The test report should present the results in a manner which is useful for the requester. To best accomplish this, the report writer should address each element of the original test request in the final test report. Specified limits and acceptance criteria should be included with the test results, so it is easy to see if the data conforms to the requirements. This approach gives direct answers to the questions

TABLE 8.3 Recommended Format for Engineering Test Reports

Front Matter
    Title page (including approvals)
    Table of contents
    List of tables
    List of illustrations

Summary
    Statement of what the test was intended to accomplish
    Brief statement of primary test conditions
    Major conclusions reached

Background or introduction

Description of test
    Test articles
    Test facilities/equipment used
    Description of test methods
    Instrumentation and measurements
    Test results
    Analysis/discussion of results
    Conclusions and recommendations

Appendix
    Detailed data sheets
    Specialized analysis (e.g., calculations)
    Test procedures (if significant to the test methods or results)

References or Bibliography

TABLE 8.4 Guidelines for Better Test Reports

Present a summary of the results for management overview.

Arrange the test results in the same sequence as the test request.

Show acceptance criteria (limits/tolerances) for each test parameter. Present the actual results adjacent to the acceptance criteria for ease of comparison.

Include detailed identification information about the items tested (part name, part number, model number, heat number, serial number, etc. for traceability of results to a particular design or material).

Describe the test methods and test setup in sufficient detail so a knowledgeable reader can judge the validity or limitations of the test results. Include photographs or sketches wherever possible.

Describe all test anomalies and failures (intentional or otherwise) and discuss significance.

Include photographs or sketches of parts tested.

---

the requester was asking. Table 8.4 presents several tips for preparing effective test reports. These should be considered when preparing such a document.

8.4.2 Review and Approval of Test Reports

After the test report is written, it should be reviewed by the management of the test organization. The purpose of this review is to verify the clarity and completeness of the report in satisfying the test request.
When this is completed, the report can then be approved and signed off by the testing group. But this is not necessarily the final approval. It is a preferred practice for design assurance purposes of having the test requester also review and approve the report. This is particularly useful if the test requester is in the product design organization. Such a practice closes the loop in the design verification process.

8.4.3 Test Records

The test records files represent one other important aspect in the control of engineering tests. These records provide objective evidence of how the design was investigated prior to its introduction into service and should be retained for future design reference. This may be of particular significance if there should be some future claims or litigation about the adequacy of the product when it is in use.

Some of the key questions about test records are: What should be kept? By whom? Where? How long? Answers to each of these questions may vary with the company, the product, and the industry. However, there appears to be several practices which can be applied with reasonable confidence in the absence of pre-designated requirements. Table 8.5 lists a suggested group of records that should be included in the test files, which answers the question of what to keep.

In regard to the question of who should keep the records, the test organization normally would be the repository for test files. Such files provide the technical backup for each final test report. Some companies prefer to have the test records transferred to the design group. This may be acceptable if the tests were subcontracted to an external testing agency, but the general rule of records management should be followed wherever possible: The group originating the document should retain the official file for it. This includes the data used in developing the document.

The matter of record retention time is not easy to answer in terms of the specific number of months or years. One guideline that may be helpful is to retain the test files for at least the warranty period or for a reasonable period of operational service. Test records are often influential in proving a company's efforts to produce a safe and reliable product. Be sure your test records provide that support for your products and are available when needed.

TABLE 8.5 Recommended Contents of Test Files

Test request and approved changes.

Test plans/procedures.

Completed test data sheets.

Photographs, drawings and sketches of the items tested and the test setup.

Description of anomalies and their resolution.

The final test report.

Other documents which contribute to the understanding and validity of the results and conclusions.

From: J. A. Burgess, Quality Assurance in Testing, Quality, Jan. 1983, Hitchcock, Wheaton, IL.

8.5 SUMMARY

The quality of test results can be improved by incorporating the various design assurance techniques described herein. Although the methods are relatively simple and straightforward, the testing organization must be methodical in the implementation of these methods. As in other chapters on design assurance topics, the test planning, test performance and test reporting requires considerable attention to detail. But this effort pays off, because it contributes directly to the validity of the final results. For design verification purposes, that's the bottom line.

9

Design Reviews

9.1   Introduction   166

9.2   Formal Design Reviews   167

    9.2.1   General Guidelines   167
    9.2.2   Planning the Design Review   169
    9.2.3   Conducting the Design Review   172
    9.2.4   Followup   177
    9.2.5   Customer Participation   178

9.3   Requirements Review   179

9.4   Design Verification Reviews   179

9.5   Informal Design Reviews   181

9.6   An Integrated Approach   181

9.7   Summary   182

9.1 INTRODUCTION

Planned and systematic design reviews are an effective means for reducing the risks associated with the introduction of new or revised products. It brings together specialists and knowledgeable generalists for the purpose of optimizing the design. Design reviews provide an opportunity for bringing questions and various points of view about a product into the open. They also allow the company to benefit from the experiences of its senior personnel. This improves the chances that a new project will avoid the problems of previous ones. In addition, the design review process allows the specialists to examine whether or not

the product was designed in accordance with the applicable requirements and to see if the proper design methods were used.

Several types of design reviews are applied in industry. Some
tend to be quite formal, whereas others are informal. Nevertheless,
each type, whether it involves one person or many, can be helpful and
has a role to play in assuring the quality of design. The most frequent-
ly-used types of design reviews include:

Formal design reviews
Requirements reviews
Design verification reviews
Informal design reviews

Each of these will be discussed in further detail in this chapter.

9.2 FORMAL DESIGN REVIEWS

As its name implies, a Formal Design Review is a structured process.
Various military and government agencies (e.g., Army/Navy/Air Force,
NASA) have developed the process that has tended to be the standard
approach for many design reviews. Broadly speaking, Formal Design
Reviews are held on a pre-established date and are conducted in the
format of a group meeting. A chairman leads the meeting and representatives from various groups review and comment on the design.
Minutes of the meeting are then published to document agreements and
items requiring further investigation or revision. The details of Formal Design Reviews are described below.

9.2.1 General Guidelines

Selecting the proper participants is one of the most important steps in
conducting an effective Formal Design Review. If the new product is
closely related to the company's historic product lines, a few experienced persons may be all that are needed. But, if the new product is
substantially different from the traditional products, or if it will
involve new or different technology, it is usually necessary to bring in
a group of technical specialists. Some companies establish a standing
committee to serve as a permanent design review panel. Others favor
the ad hoc committee approach, picking certain persons from various
groups in the company, based on the particular needs for each design
review. Still other firms simply invite each department to send one or
two representatives to participate. Each approach will work if the necessary planning and preparation is accomplished.

It is common to seek participants who are knowledgeable of the disciplines most commonly found in industry - design, manufacturing,
marketing, purchasing, quality control and field service. Depending
upon the product, it may also be appropriate to include other special-

ists, such as safety, stress analysis, maintainability, etc. The guiding principle is to invite only those who can make a significant contribution. Also it is a good practice to limit the total number of participants to 5-10 persons. For these reasons, the ad hoc review team of specifically-selected persons is generally preferable.

A chairman should be appointed to lead each design review. Normally, the chief engineer or project manager will appoint someone from the Engineering Department to serve as chairman. The person selected should have a broad understanding of the company's products and a solid technical background. The chairman should also be skilled in leading technical meetings and maintaining control tactfully.

For small companies with only a few people who have technical expertise, it is sometimes necessary to seek help for the outside to get the necessary level of objectivity. Local consulting engineers or retired technical people can be helpful in this capacity. They can assist the small company in avoiding tunnel vision about its products and approaches caused by inbreeding.

One aspect of staffing the design review group is especially important. The person selected to be the design review chairman should not have direct responsibility for the design being reviewed. In addition, the chairman should not be in the line of authority over or under the designer. The chairman must be able to lead the review through a thorough investigation of the design, without limitations imposed by direct involvement or self-interest.

All participants, including the chairman, must be given the opportunity to review the design package prior to the design review. Otherwise, their participation will be much less effective. Time spent in preparation greatly increases the effectiveness of the process.

Each Formal Design Review should be documented by a report or meeting minutes. Some firms issue a report, consisting of the review package (drawings, specs, worksheets, etc.) plus the compilation of agreements and assigned action items. Other companies simply write a set of meeting minutes which identify the portions of the design which were considered acceptable and those where questions or concerns must be resolved.

It should be the chairman's responsibility to see that a useful record of the design review is made. He may choose to do it himself or appoint another participant to act as secretary for the design review. In large companies, a reliability engineer is frequently given this assignment.

When should a design review be held? The answer to this is not simple. It depends upon a lot of factors, such as the product, the market situation, the production cycle, and company size to name a few. Nevertheless, the intent of the design review is to minimize the risk of introducing change. Therefore, the review must be held in a timely fashion so the new or revised product can realize the benefits from the process in a cost and quality-effective manner. In fact, it is generally appropriate to hold more than one review during the development and introduction of the product. Figure 9.1 illustrates a time frame for conducting design reviews throughout the development and production cycle. In addition, Table 9.1 describes the various types of Formal Design Reviews shown in Figure 9.1.

And finally, what does the design review process cost? One NASA study several years ago estimated that design reviews on several space projects cost about one to two percent of the total engineering cost. A prominent machinery company figures a major review costs about $25,000 and a minor review about $10,000.

For small design projects, consider that six-eight persons for two or three days at $200 a day would cost the company $2,500-$5,000. Or a consultant for ten working days at $500 a day plus expenses might cost $6,000-$8,000. These figures are small when compared to the cost of a product recall, a damage suit or a market flop that might have been avoided by a comprehensive design review.

It must be remembered that design reviews are not a panacea for all problems. Poor selection of participants, inadequate preparation and lack of follow-through can severely impair the potential that a good design review offers.

The material presented in the next sections provide guidelines and recommendations for achieving the benefits of an effective Formal Design Review.

9.2.2 Planning the Design Review

Leadership for design review planning typically rests with the engineering management of the company. The chief engineer or department manager should establish a policy in regard to Formal Design Reviews. Under what circumstances should a review be conducted? How many are needed? Where in the life cycle? Then as the appropriate situations occur, engineering management should select a design review date at least a month or more in advance and announce it to the organization. The chairman should be appointed at that time, and all affected parties should be advised of the objectives and scope of the review. The burden of the task then shifts to the design group or the individual designer to prepare for the review.

The person responsible for the particular design then must gather together the necessary specifications, drawings and related material which describes the design and the approaches taken. This information must then be reproduced and distributed to the appointed reviewers. Table 9.2 suggests items to include in the review package. The information must be sent out sufficiently in advance to give the participants time to become familiar with the design and the requirements. Generally, two weeks in advance of the review meeting is a good rule

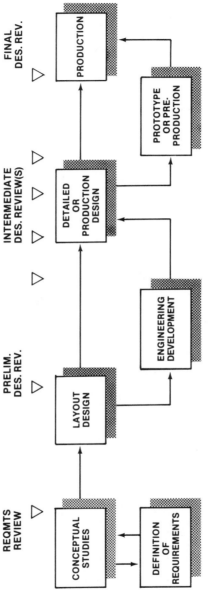

FIGURE 9.1 Suggested points for formal design reviews.

TABLE 9.1 Types of Formal Design Reviews

Preliminary design review:	A review of the basic design concept and an evaluation of the concept against the preliminary design specification. The purpose is to examine the early design for compliance with the overall design requirements.
Requirements review:	A review specifically aimed at examining the appropriateness and completeness of the design requirements for a particular component or project.
	The Requirements Review is occasionally performed in conjunction with the Preliminary Design Review.
Intermediate design review:	A review conducted after the layout design is complete. It is normally held before the production drawings are prepared. The purpose is to evaluate the design against the detailed requirements. For major projects, more than one Intermediate Design Review may be needed.
Final design review:	A review performed after at least one prototype or pre-production unit has been built and tested. The purpose of this review is to fine-tune the design prior to authorizing full-scale production and release for operational use. The review focuses on accomplishment of performance, cost, producibility and reliability objectives.

TABLE 9.2 Typical Contents of a Design Review Package

Layouts, schematics or design drawings

Design specification or requirements sheets

Summaries of available calculations or test results

Performance and reliability analysis data

Cost estimates and projections

Description of unusual requirements or high-risk elements
in the project

From John A. Burgess, Making the Most of Design
Reviews, *Machine Design*, Penton/IPC, Cleveland,
Ohio, July 4, 1968, pp. 90-95

of thumb. Failure to distribute the review package or sending it out
without allowing time for review will result in the review meeting being
an education session for the reviewers, rather than being a session
where the reviewers provide useful insights to the designers.

In parallel with the engineering department preparation for the design review, the chairman is responsible for working out the various
administrative details, such as selecting a meeting place, developing
the agenda, and handling the logistics of the meeting (chairs, projec-
tors, note taking etc.). Tables 9.3 and 9.4 list many of the items that
should be considered.

When developing the design review schedule, the chairman should
establish reasonable time constraints. Most persons have an interest
span limit of about four hours on any given topic. Fortunately, that
should be long enough to cover most projects. Many can be handled
in a couple of hours if the reviewers have come prepared. If it is
necessary to go beyond approximately a half-day meeting, it is preferable to divide it into two or more review meetings of two to four hours
each. However, if this is necessary, it is very likely that a different
group of reviewers may be needed for each separate session. The
guiding principle is do what has to be done, but don't overkill it.

9.2.3 Conducting the Design Review

An effective design review format consists of three distinct phases.
The first is an introduction to the design, the second is the review
discussion, and the third is the wrap-up. Each of these are described
below.

TABLE 9.3 A Design Review Planning Schedule

Activity	Work days from design review
Schedule design review	D-30
Publish agenda. Assign personnel to topics. Invite review board members.	D-25
Initial illustrations available. Send out available review packages to review board members.	D-10
Dry runs	D-7 to D-3
Final illustrations available	D-2
Final dry run	D-1
Design review	D-Day
Critique	D+1
Issue design review summary report	D+10

From John A. Burgess, Making the Most of Design
Reviews, *Machine Design*, Penton/IPC, Cleveland,
Ohio, July 4, 1968, pp. 90-95.

The introduction phase should begin with a brief statement by the
Design Review chairman. The chairman should set an objective tone for
the meeting. He should state the purpose of the review and emphasize
the need for objectivity and constructive inputs by all participants.
Table 9.5 presents several guidelines for the chairman to use in leading
the meeting.

After the chairman concludes his introductory remarks, he should
then turn the meeting over to the responsible engineer to introduce the
design.

This portion of the introduction should provide the reviewers with
the big picture and establish the similarity and familiarity to the company's existing products and capabilities. Table 9.6 contains a number
of recommendations for items which should be included in the introduction. These, of course, must be tailored to the local situation. Approximately 10% of the total allotted time should be spent on the introduction.

Next, the designer should proceed with the second phase of the

TABLE 9.4 Chairman's Last-Minute Checklist

Sufficient chairs available and seating arrangement checked

Room ventilation and air conditioning settings checked

Display material available

Security measures checked

Chalk and eraser available for chalkboard

Projection equipment and screen available (including spare projection bulbs)

Trained projectionist available

Receptionist alerted if persons from outside the company are invited

Arrangements made for coffee breaks

Pencils and paper available for note-taking

Persons assigned to take minutes (record action items, questions and significant comments)

Ash trays available

Podium, pointers, public-address system available if needed

Persons available to operate light (or window shades) if room is to be darkened

---

From John A. Burgess, Making the Most of Design Reviews,
Machine Designs, Penton/IPC, Cleveland, Ohio, July 4,
1968, pp. 90-95.

review—a detailed discussion of the new product. This discussion should proceed methodically and cover the various aspects of it. This includes a brief description of each significant requirement and how it has been satisfied or accommodated in the design. Usually, the reviewers will ask questions as the presentation is being made. This will include inquiries about alternate approaches and why the particular design was chosen. The designer and other responsible engineering personnel should anticipate these and similar questions and be prepared to answer them.

Another area of considerable interest to an experienced reviewer are the assumptions used in the design process and the areas of risk or uncertainty. It is this aspect, in particular, that the experience and judgment of the reviewers is brought into play.

However, a word of caution is needed at this point. All persons

TABLE 9.5 Design Review Leadership Techniques

Start on time, break on time, stop on time.

Introduce outsiders and newcomers.

Make sure everyone can see and hear what is going on.

Start the meeting by stating the objectives of the meeting. (Stress the fact that Design Review purpose is to help the designer, not criticize.)

Make sure that minutes are taken.

Keep the discussion on the track.

Encourage participation by everyone but discourage debates.

Ask searching questions to assure the group has considered all aspects of the subject.

Maintain impartiality. Avoid giving your own personal opinion.

Sum up periodically to keep the review moving forward.

Keep firm control, but do not be a dictator.

At the end of the meeting, summarize the conclusions. Identify what action must be taken, when, and by whom.

From John A. Burgess, Making the Most of Design Reviews, *Machine Design*, Penton/IPC, Cleveland, Ohio, July 4, 1968.

TABLE 9.6 Suggested Introductory Topics

Background data for overall orientation

Events or circumstances leading to this project or design

Summary of significant design requirements

General design approach (similarities and differences from existing product lines)

New features which are intended to overcome or avoid previous problems

Key assumptions used in the design

Alternative designs considered

Selected approach and basis for selection over alternates

Problems encountered or risks anticipated

---

participating in a design review must recognize and remember why they are there at all. It is to make a constructive review of the design and minimize the risks of introducing change. It is not to attack the designer or criticize the Engineering Department. Maintaining an atmosphere of objectivity and professionalism is a key duty of the chairman. Don't forget, designers typically are very proud of their work and may become very defensive when they or their designs are criticized. The chairman must retain sufficient control of the meeting and keep the discussion process on target.

The designer should be prepared to explain the important methods used to demonstrate the design will meet the applicable requirements. This includes showing the results from design calculations, development tests or experience gained from models or trial demonstrations. Several of the engineering tools described in earlier chapters, such as design checklists (Section 2.5) analysis reports (Section 6.3), and test reports (Section 8.4) provide objective evidence for supporting the design.

An important tool for the reviewers to use in this portion of the review meeting is the Design Review Checklist. The checklist is a memory-jogger of many different factors which should be examined in a review process. Although it is most useful to tailor a design review checklist to a particular product line or industry, various general checklists exist in industry and can serve as a basic framework for local tailoring. Checklists are not meant to limit the investigation, but they do reduce the chances of oversight. One such Design Review Chcklist is shown in Appendix 2. Even though each reviewer is responsible for evaluating the design in light of their own experience and area of expertise, the chairman should at least use a generalized checklist to assure all important aspects of the design are reviewed. In particular, the chairman should be sure that the areas of lesser visibility but of significant importance (e.g., operating instructions, special tools, warning labels, spare parts, packaging/handling, etc.) are covered.

The team of reviewers should examine the data presented and conclusions reached by the designer. They should consider the following questions:

Do the data and results support the conclusions drawn?
Do the assumptions seem reasonable?
Are there areas where the risks appear to be higher than normal?
Are there items of significance that have not been addressed?
Does the design satisfy the applicable requirements? Are there
    requirements which were intentionally not satisfied? Is that
    acceptable?
Are the design methods used appropriate for this product and its
    intended application?

What problems remain to be solved? Is there adequate assurance
the problems can be solved in a reasonable manner and appropriate time frame?
And, finally, does the design appear to be satisfactory to proceed
to the next phase of development or production?

Answers to these questions, and similar ones, are an important
output from a Formal Design Review. The chairman must make a point
to see these kinds of questions are asked and answered during the
final, or wrap-up, portion of the review meeting. It is a good practice for the chairman to poll the reviewers individually to solicit their
concerns and to identify items requiring further action. The resulting
list of open items should be recorded in the design review documentation.

9.2.4 Followup

The key ingredient of successful followup is the resolution of open
items identified in the design review. At least two different approaches
are frequently used in industry today for handling this. One approach
is to assign the responsibility for resolving the action items to the design review chairman. Even though he normally returns to his regular
duties, he is expected to track and obtain final closeout of the concerns and action items. This approach retains the original objectivity
of the chairman in getting satisfactory resolution. However, it tends
to result in a slow turnaround time for reaching closure, since interest
and involvement tends to wane with the passage of time.

The other common approach is to assign the responsibility to the
Engineering manager. Since the product design is the responsibility
of the Engineering Department, it is natural to have them complete the
task by resolving the open items. Some persons outside of Engineering
express the fear that Engineering will not give the concerns proper
consideration. However, there is seldom any real basis for such
worries. In fact, the second approach usually results in the quickest
resolution of the problems and is preferred in most companies.

A word of caution about design review open items—it can be both
embarrassing and costly to find that an identified problem was ignored
and later caused production delays, expensive repairs or customer
complaints. Don't treat the concerns of reviewers lightly. Remember,
they were invited to participate because someone thought they had
special knowledge or experience to contribute.

From the standpoint of assuring quality in design, each documented action item should be investigated objectively and its resolution
documented as part of the design review followup. In some cases,
this will result in an actual change in the design. In other cases, it
will simply mean conducting further tests or calculations, which

strengthen the earlier conclusions. And in other cases, it will result in an explanation, supported by technical data, which shows the original concern was unfounded. Nevertheless, each item deserves evaluation and a response to the review team to close it out. Anything less is incomplete engineering work.

There is one other element of followup to consider. Some firms effectively use a critique process after a design review. The purpose of the critique is to examine what worked well and where improvements in the process can or should be made. This is best accomplished by holding a short meeting within two or three days after the Formal Design Review is completed. Persons involved in the critique should include the Engineering Manager, the Design Review Chairman and one or two senior persons who were in the review. This serves as a postmortem of the approach. It is very useful in sharpening the skills of the engineering organization in their use of design reviews.

9.2.5 Customer Participation

Many government contracts include requirements for design reviews. The purpose of these reviews is to assess whether the project is ready to proceed to the next phase. Representatives from the governing agency serve as the review team. Their concurrence is normally needed before the contractor can move ahead. Thus, it is imperative for the Engineering organization to do a thorough job in preparing for the review.

The basic points presented in this chapter on Formal Design Reviews apply also to a Customer Design Review. The major differences are in the preparation and followup phases.

From the standpoint of wanting to put your best foot forward, it is a good idea to conduct a dry run or an internal design review a few days before the customer review meeting. This gives company management the opportunity to examine the planned presentation of the design prior to direct customer involvement.

For maximum benefit, the dry run should be conducted using the drawings, charts, illustrations, etc. in the final form that will be used in the customer review. A few senior people from the management staff can serve as the internal review team for the dry run. Their job is to ask questions that they anticipate the customer representatives will ask and to look for weaknesses in the design approach or unsupported conclusions in the technical presentation.

Engineering management can then take some last minute action to resolve any weaknesses discovered or at least have a plan for corrective action. Otherwise, these same problems discovered during the Customer Design Review could have a delaying effect on obtaining customer concurrence to move ahead to the next phase of the project.

The other aspect of the Customer Design Review that is different from conventional in-house design reviews is the follow-up phase.

The customer routinely will expect to receive meeting minutes or a written report documenting the design review process. This will usually have to be furnished within a month after the review. The customer typically will expect to receive an action plan and schedule in the report, covering the resolution of the open items identified during during the customer review meeting. This will require the company's management to be prepared to investigate the action items thoroughly and to make commitments to resolve them. Failure to do this in a timely and comprehensive manner may cause delays in obtaining customer approval to proceed. It is an area that must get management attention from the start.

9.3 REQUIREMENTS REVIEW

A Requirements Review is a special type of design review. It has many applications; unfortunately, it is frequently omitted. The purpose of a Requirements Review is to obtain assurance that the proper set of requirements has been defined and is now available for use in the design and development process.

A Requirements Review is similar to a Formal Design Review. A group of knowledgeable persons are gathered to review and concur with the designer's selection of requirements. The designer prepares an initial compilation of requirements in accordance with the concepts and methods discussed in Chapter 2. Then the resulting design requirements document is circulated to the review team for examination prior to the meeting. The reviewers submit their comments to the designer and then meet at a designated time and location to discuss the comments.

Although many firms use the review and comment technique, many do not follow through to resolve differences among the reviewers or between the designer and a reviewer. A resolution meeting serves as an effective means for closing the loop and for getting a better understanding of why certain requirements must be included. It also is a good tool for integrating the efforts of various disciplines. This is achieved through the interactions among the specialists and the designer.

The most common output of a Requirements Review is a design specification which then becomes the foundation on which the product design is built. Some companies have the review team members sign off the final design specification to show concurrence. Although this is not mandatory, it can be helpful.

9.4 DESIGN VERIFICATION REVIEWS

The purpose of a Design Verification Review is to provide assurance that detailed design efforts are correct and proper. Although a Formal Design Review can be used for design verification purposes, many of them tend to focus more on systems, large components and overall product design adequacy. Participants in a Formal Design Review are only mildly interested in the mathematical accuracy of a calculation as it effects a conclusion or supports a decision. However, a Design Verification Review focuses on the day-to-day design effort to assure the correct design method was used and the results are accurate.

Design Verification Reviews are frequently conducted by one or two persons who are knowledgeable of the detailed design requirements and methods and have access to the applicable design information. A Design Verification Review typically does not use a group meeting format, and no one is designated as the chairman.

For greater effectiveness, the verifier should be another qualified designer who is capable and experienced in doing this design work but had no direct responsibility for it. Many companies use the designer's supervisor in this capacity. This approach is workable, but some do not favor it. The critics say the supervisor lacks the necessary objectivity due to the high degree of involvement in the original design process. The other often-voiced criticism is that supervisors have many demands upon their time and are inclined to perform only a cursory technical review.

Regardless of who the verifier is, the duties should be the same. The verifier is expected to examine the specific assumptions, design methods and the detailed elements of the design process. It is the verifier's responsibility to check the work in sufficient detail to be comfortable with the results. This may be from direct experience with similar designs, performing alternate, simplified calculations to check the order of magnitude of the results, or checking portions of the specific work to determine it is correct and accurate.

When the verifier is satisfied that the design is proper and accurate for the specific application, the verifier is normally expected to sign one or more of the design documents, signifying his concurrence. This process is intended to reduce the chances of error or oversight by the original designer. It is a form of technical inspection of the engineer's work. Although designers may cast a leary eye on someone routinely checking their work, there is a substantial body of experience that shows the Design Verification Review usually pays off in fewer errors and later design changes. Government contracts for high-technology programs frequently require some form of design verification reviews. For example, companies involved with the design of nuclear power plants and related equipment routinely apply Design Verification Reviews, usually performed by another qualified person in that technical specialty (pump designer, pressure vessel designer, electrical controls designer, etc.). It is a very basic method for minimizing errors at the detail level.

9.5 INFORMAL DESIGN REVIEWS

Another useful approach is the use of informal design reviews. This process lacks most of the structure of the Formal Review (no chairman; no single group meeting; little, if any, documentation). It is often accomplished in a one-on-one setting with occasional small group gatherings around the drawing board or a chalkboard. Individual specialists assigned to the project from manufacturing, quality control, materials and other functional groups periodically drop by to see how a new or revised design is shaping up. In the process, they provide technical inputs or answer the designer's questions in their area of expertise. The approach is especially effective in building a rapport between individuals, and it largely avoids the designer's uncomfortable feelings that have been known to occur in a structured design review meeting. The success of the informal approach, of course, depends heavily upon the reviewers and the designer to put enough time and effort into the informal reviews. However, these unstructured technical exchanges can be very beneficial to the designer in reducing the risks associated with the new design.

The informal review is frequently tied in with a drawing/specification signoff procedure. Where the designer and the reviewers learn to work together, the signoffs normally are accomplished routinely and without delaying the design cycle. From a design assurance standpoint, it is a cost-effective approach for achieving a high level of quality in the final design.

9.6 AN INTEGRATED APPROACH

Small companies with relatively uncomplicated products may not need much of a design review effort or may not have the resources to accomplish it. However, medium to large-sized companies with numerous product lines and companies with high-technology products, regardless of size, can benefit from an integrated design review effort.

Such a program might consist of a published policy to conduct one or more Formal Design Reviews for each new product development or major product change. It would be natural to include a Requirements Review in the formulation of requirements. Later a Formal Design Review would be performed when the basic design is completed (for example, after the final design concept is chosen and the general configuration established). Informal design reviews would then be used to work out the detailed designs at the component and subassembly levels. Following that, multi-discipline signoffs would be incorporated for approving the drawings and specifications.

Another Formal Design Review would be conducted either just before or immediately after the construction of a prototype or first pro-

duction model. This design review would focus on the performance
and producibility of the final production design. It should normally
be held prior to the official start of production quantities.

In addition, selected individuals would be designated as design
verifiers and would review the detailed design and analysis work as it
is completed, in an attempt to achieve an error-free design.

Even though the concepts presented in this chapter have been directed at hardware-type products, the same concepts apply equally
well to service or software projects. In the past, this aspect has
largely been ignored. Nevertheless, design reviews offer the same
benefits to software and technical service companies as to the traditional hardware manufacturers.

9.7 SUMMARY

A design review is a planned and systematic study of a design by knowledgeable technical specialists. The purpose is to assure that the new
or improved product can be introduced with the least risk. Participants
in a design review examine the design and the methods used. Their
objective is to judge if the product has a high probability of satisfying
the applicable requirements, while meeting the necessary performance,
cost, reliability and safety goals.

Several types of design reviews are used in industry. The Formal
Design Review, with its structured approach, is the best known type.
However, informal design reviews can also be very effective.

Requirements Reviews and Design Verification Reviews are special
types of design reviews. They are intended to focus sharply on details and they strive for accuracy and completeness at the "nut and
bolt" level whereas the Formal Design Reviews are often directed at the
system or major component level.

Each of the design review processes can make a significant contribution towards achieving quality in design. And the processes can
also be used for software or services as well as for equipment and
hardware.

10

Statistical Tools for Design Assurance

10.1 Introduction   183

10.2 Frequency Distributions   184

10.3 The Normal Curve   188

10.4 Process Capability   192

10.5 Statistical Tolerancing   195

10.6 Summary   197

10.1 INTRODUCTION

Variation is present in nature and in all man-made items. No two things are *exactly* alike. The differences may be large or small, but nevertheless there are differences. These differences can have important effects on the product and on the quality of design. Since the impact of variability is different from item to item and product to product, it is necessary to find means for assessing its significance.

The field of statistics provides many techniques which are useful in the study of variability and its effect on design. Although some are quite sophisticated and complex, there are several simple, but powerful, tools the designer can use frequently in the day-to-day engineering tasks.

The key building block of statistics is the study of samples to estimate the characteristics of a larger population. By following certain rules and practices, analysis of relatively small amounts of data can reveal much about a product or process with a high level of confidence.

In this chapter the statistical methods will be presented and explained in very basic terms. Several excellent texts on statistics are available which go into the derivation and in-depth explanation of the theorems for those who wish to pursue this. However, it is not necessary to be a statistician to use the tools presented in this chapter. To many readers, this section will be more of a memory jogger, or thought provoker, than a revelation of new concepts. It is in this vein that it is presented.

10.2 FREQUENCY DISTRIBUTIONS

One of the easiest tools to use is the frequency distribution. It is a graphical method for displaying the data. The most common format is to construct a tally sheet as shown in Figure 10.1. It shows the number of times each measurement value occurs and the pattern it follows.

For ease of construction and clarity in interpretation of the data,

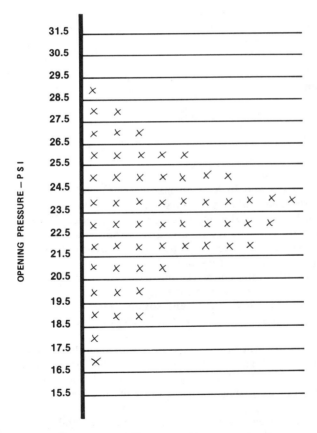

FIGURE 10.1 An example of a tally sheet.

divide the tally sheet into equal divisions, or cells. A total of 7-12
cells is convenient to work with, but it is not unusual to use up to 20
or so cells where increased visibility is needed and a large number of
data points is planned.

Each cell boundary should be marked in a manner to make each
cell mutually exclusive, i.e., there will be only one cell where the
measured value fits. The easiest way to do this is to select cell boundary values which are half the value of the measurements to be recorded.
For instance, if you were gathering data on spring force to the nearest
ten pounds, the cell boundaries should be: 5, 15, 25, 35, etc. Otherwise, if the boundaries were 10, 20, 30, etc. a value of 20 pounds could
be recorded in cell 10-20 or 20-30.

Each individual measurement is then recorded as a check mark or
"X" in the appropriate cell. A minimum of 50 measurements or values
should be plotted to be statistically valid. It is seldom necessary to
use more than 100-200 data points.

If the range of measurements is too great for the initial selection
of cell divisions, change the divisions to cover larger values but continue to use about 10-20 cells. A smaller number of cells may not provide enough discrimination among the values to show patterns. Much
more than about 20 cells may dilute any trends present, plus be clumsy
to construct.

The tally sheet format is particularly useful, because it shows
several important features of the data. By inspection, the user can
see the apparent average value, or central tendency, of the data and
the spread, or range, of the values. By drawing lines on the tally
sheet to show the upper and lower specification values, you can also
see how well the data meets the specified limits.

Although 50-100 data points are recommended, the mere process of
plotting the data in this fashion can be very beneficial. In fact, the
pattern may show what action is needed even if the number of data
points is less than the preferred minimum quantity of 50.

For presentation purposes, tally sheets are often changed into
frequency histograms. To do this, the axis is rotated from vertical to
horizontal, and the data is shown as adjacent bars, whose height is
proportional to the number of check marks in each cell. Figure 10.2
shows frequency histograms of three different conditions which are
often encountered in a engineering or manufacturing situation. Figure
10.2a is an example of a process which is well within the specification
limits and nicely centered. Figure 10.2b shows a process with a range
of variability, which is tighter than the specification requires, but the
process needs to be adjusted slightly to center it within the specification limits. On the other hand, Figure 10.2c illustrates a process
which is well centered, but the variability exceeds the allowable limits,
thus indicating a problem with the process or with the specification
requirements.

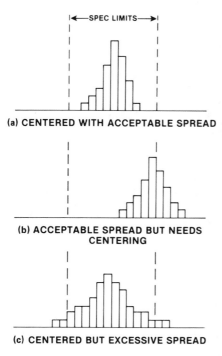

(a) CENTERED WITH ACCEPTABLE SPREAD

(b) ACCEPTABLE SPREAD BUT NEEDS
    CENTERING

(c) CENTERED BUT EXCESSIVE SPREAD

FIGURE 10.2 Examples of various frequency distributions in relation to the specification limits.

Now let's see what else we can learn from frequency distributions. Note that the frequency distributions in Figures 10.1 and 10.2 have a common shape. Each has a single peak near its center with lesser values falling off on each side of the peak value. Such a pattern is common and is an indicator that the process or activity is operating in a reasonably normal and consistent manner. However, other patterns are sometimes found that are a signal to look harder and longer at the data before taking action on it.

Figure 10.3 shows several other frequency distributions which should be recognized as special situations. The histograms in Figures 10.3a and 10.3b represent a skewed, or biased, distribution. This could be caused by accident or by intent. For example, a new lathe operator might intentionally make shafts on the high side of the diameter tolerance to avoid making any undersized parts. On the other hand, a machine stop might have come loose and allowed the machine to drift in one direction but not the other. Skewed distributions nearly always are due to an assignable cause and are not representative of a chance occurrence.

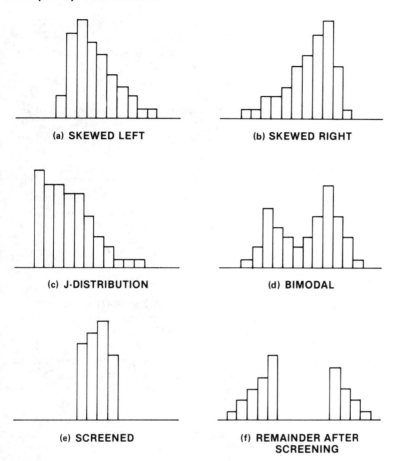

FIGURE 10.3 Examples of other types of frequency distributions.

Figure 10.3c is called a J-distribution and is often associated with a situation where there is a natural limit, such as a zero value. Again, such a distribution can generally be traced to an assignable cause.

Figure 10.3d illustrates a bimodal distribution. Notice there are two (or more) peaks which are separated by values substantially lower than the peak values. This situation is nearly always indicative of a mixing of two (or more) batches of data. It might be from different tests, different operators, different suppliers, etc. Before any meaningful decisions can be made or actions taken, it is necessary to separate the different batches of data and examine each one individually.

Figures 10.3e and 10.3f represent a cause and effect situation. In Figure 10.3e, the histogram shows a process where values beyond the upper and lower limits were screened out, and only items within the

limits remain. This is representative of corrective action taken on a
process such as shown earlier in Figure 10.2c.

Figure 10.3f shows the pattern of what remains after the screening
operation.

These patterns and others which deviate significantly from the preferred distribution shown in Figure 10.2a are warning flags that something unusual is, or has, taken place. Thus, further investigation is
warranted.

Tally sheets and frequency distributions have many applications in
design, manufacturing, testing, data analysis and troubleshooting.
They are easy to construct and easy to interpret. No mathematics are
required and non-technical people are quick to understand the message
they convey. Often the data is already available, and it's only a matter of gathering and plotting it. In other cases, you can frequently
get other persons to record the information in this form as the operations are performed or measurements made.

10.3 THE NORMAL CURVE

Operations found in industry tend to follow a consistent pattern.
Values, whether dimensions, voltage, weights, horsepower, etc., tend
to be grouped around the specified value. Thinking in terms of the
frequency histogram, the largest number of occurrences tend to be
near the specified value and fewer and fewer occurrences are found as
you move away on either side. This is referred to in the statistical
literature as a binominal distribution, but it is more commonly called the
Normal Curve. Figure 10.4 illustrates how the binominal distribution
is enclosed by a symmetrical curve.

In actual practice, many situations occur in industry that are close
approximations to a binominal distribution. Even though the samples
may not plot perfectly, the data population is usually close enough to
allow using the Normal Curve relationships for analysis purposes.

The Normal Curve has some very useful characteristics. The first
is its central tendency, or average value, defined as X-bar ($\bar{X}$). This
is the most commonly-occuring value in the data. It is computed as
the weighted average:

$$\bar{X} = \frac{X_1 + X_2 + X_3 + \dots X_n}{n}$$

The second important parameter of the normal curve is its spread,
or dispersion. This is most commonly measured in terms of its range
(R) or the standard deviation, defined by the lower case Greek letter
sigma (σ). The formula for range is:

R = Max. value - Min. value

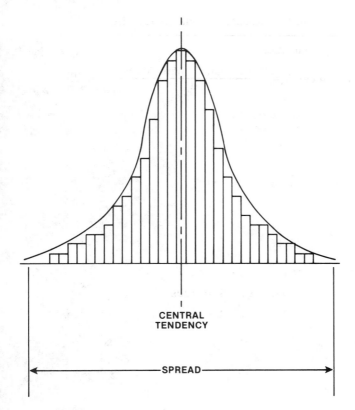

FIGURE 10.4 Normal curve and a binominal distribution.

The formula for standard deviation is more complex as shown below.

$$\sigma = \sqrt{\frac{(\overline{X} - X_1)^2 + (\overline{X} - X_2)^2 + (\overline{X} - X_3)^2 + \ldots (\overline{X} - X_n)^2}{n}}$$

The range is less precise as a measure of dispersion than the standard deviation. However, it is still useful in providing good approximations of the spread. Table 10.1 describes two methods for estimating the standard deviation of small numbers of samples, using the range of the data. As is obvious from the equations, long-hand calculations of averages and standard deviations can be tedious. However, several hand-held calculators and commercially available computer software programs can be used to make the calculations quite painless.

The third important characteristic of the Normal Curve is the way the area under the curve is distributed. Figure 10.5 shows this

TABLE 10.1 Using Range to Estimate Standard Deviation

Method A (Close Approximation)

$$\text{Est. Std. Dev. } (\sigma \text{ est.}) = \frac{\text{Max. Value - Min. Value}}{d_2}$$

Sample Size	Factor d_2
3	1.69
4	2.06
5	2.33
6	2.53
7	2.70
8	2.85
9	2.97
10	3.08

Example: Fracture loads for six beams

      910 lbs.   950 lbs.   1,040 lbs.

      870 lbs.   1,020 lbs.   1,010 lbs.

$$\sigma \text{ est.} = \frac{1040 - 870}{d_2 \,(6 \text{ samples})} = \frac{170}{2.53}$$

σ est. =   67.2 lbs.

Method B (Rule of thumb)

For sample sizes 4 to 9:

$$\text{Est. Std. Dev. } (\sigma \text{ est.}) = \frac{\text{Max. Value - Min. Value}}{\sqrt{\text{Sample Size}}}$$

Example: Fracture load for six beams
(same data as above)

$$\sigma \text{ est.} = \frac{170}{\sqrt{6}} = \frac{170}{2.449}$$

σ est. = 69.4 lbs.

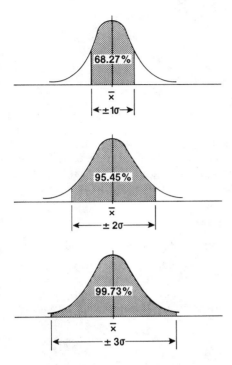

FIGURE 10.5 Areas under the normal curve.

distribution in terms of the distance (in multiples of one standard deviation) from the central tendency ($\overline{X}$). This characteristic is especially useful in studying variability.

You may ask, "What does this have to do with my design effort or
with design assurance?" By using the statistical approach, the designer can assess the extent that variability enters into product design. Variability exists in the dimensions of parts, properties of
materials, applied loads, environmental conditions, coefficients of
friction, rate of heat transfer, etc., etc. These and many other variables are present in different degrees and in different amounts during
the life and use of the product. By knowing how to use the characteristics of the Normal Curve, the designer can assess the degree of
variability present in test or operating data.

The example below illustrates one of the several ways this phenon-
enon can be used.

A designer of motor-operated valves received an order which
specifies that the valve must open from the closed position in
4.0-5.0 seconds. Test results from the first six valves showed
opening times of:

4.3 seconds
4.7   "
4.0   "
4.8   "
4.9   "
4.8   "

The engineer noted that all values were within the customer's
tolerance but decided to investigate the extent of variability
present. He calculated the average ($\overline{X}$) as 4.58 seconds.
Using a hand calculator, he also calculated the standard deviation
as 0.35 seconds.* With this information, he realized that due to
normal variability in the design and manufacturing processes, the
opening times could range from:

$\overline{X} + 3\sigma$; 4.58 + 3 (.35) = 5.63 seconds

$\overline{X} - 3\sigma$; 4.58 - 3 (.35) = 3.53 seconds

This shows how the designer could have been misled to think the
design met the requirements if he would have based his decision only
on the individual test results, without considering the effects of the
variability present in the total (including future) population of valves.

10.4 PROCESS CAPABILITY

Another useful skill for the designer is to know how to assess the capability of a machine or process to meet the specified requirements.
Process capability studies use the results from actual production or
process output to evaluate the variability inherent in the process.
This knowledge is then useful in determining how realistic are the
product tolerances or whether or not new equipment may be needed to
meet the requirements.

Two different methods are frequently used. The first is a graphical approach, the other a mathematical method. In each case, the
characteristic of interest is selected, and a series of data points are
compiled. For large volume items, 100-200 data points should be used.
For low volume items, use at least 25 data points but 50 are preferable.
Successive items are measured and the individual results are plotted
on a chart which also shows the upper and lower specification limits.
See Figure 10.6 for an example. If it is noted initially that some data
points are falling outside the limits, adjustments to the process or
machine settings can be made to see if the output can be brought into
tolerance. However, it is necessary to compile the recommended number

---

*When using less than 10-15 data points, it is a conservative practice
to calculate the standard deviation using (n-1) instead of (n).

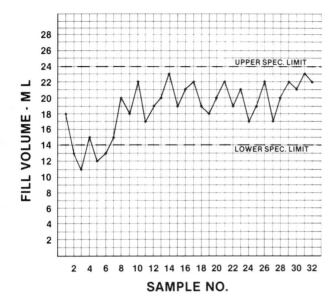

FIGURE 10.6 Graphical plot to examine process capability. (Note the shift in the data as a result of a process adjustment.)

of successive measurements after the last adjustment is made. By plotting the final batch of data, you can get a good approximation of the process capability simply by examining the data pattern in comparison to the limits.

A more exact method is to calculate the average ($\bar{X}$) and the standard deviation (σ) for the final 50-100 measurements (those taken after the last adjustment to the machine or process). Then calculate the $\pm 3\sigma$ spread around the average ($\bar{X}$) and compare that to the specification tolerance.

If the 6σ value is less than the allowable specification value, as shown in Figure 10.7a, then the machine or process is compatible with the specification limits. However, if the 6σ value is greater than the specification limits as in Figure 10.7b, the process is not capable of consistently holding the specified tolerances. Only three courses of action are then available.

1. Change the specification
2. Change the process
3. Live with the situation and sort conforming from non-conforming

This approach is a useful technique to use before deciding to arbitrarily relax an established tolerance or recommend investing in

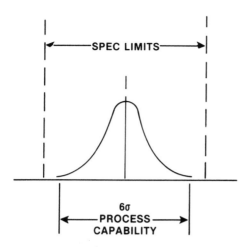

(a) WITHIN SPECIFICATION LIMITS

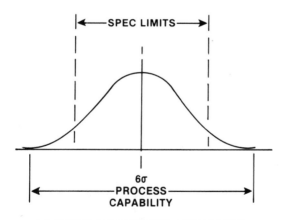

(b) OUTSIDE OF SPECIFICATION LIMITS

FIGURE 10.7 Process capability vs. specification allowances.

an expensive new machine. A little bit of data can be very revealing when examined in this manner.

10.5 STATISTICAL TOLERANCING

Another statistical tool for the designer to use is statistical tolerancing. Wherever two or more parts fit together, there is a range of
values from the "max-max" to the "min-min" conditions. (All items
maximum dimension or all items minimum dimension). Although this
approach is mathematically correct, in everyday practices, the two
extreme conditions are seldom encountered. For instance if two
mating parts (such as a shaft and a sleeve) are produced in quantity production by reasonably stable processes, only one percent of
the shafts will be at the maximum diameter and only about one percent of the sleeves will be at the minimum diameter. As a result,
there is a very small probability of the extremes being randomly
selected for use together:

$$\frac{1}{100} \times \frac{1}{100} = \frac{1}{10,000}$$

For high-volume production assemblies, it is often too conservative to use the maximum-minimum stackup of tolerances. This can
result in excessive tolerance in the final assembly or unusually-tight
tolerances for the components.

An alternate approach is to achieve reasonable final assembly
tolerances by applying statistical tolerances according to the square
root of the sum of the squares. These are applied as:

$$T_A = \sqrt{(T_1)^2 + (T_2)^2 + \dots (T_n)^2}$$

Where

T_A = Tolerance of the assembly

T_1 = Tolerance of component 1

T_2 = Tolerance of component 2

T_n = Tolerance of the nth component

Figure 10.8 shows an example of this. Statistical tolerancing can
be used when the following conditions apply:

1. All of the parts are produced independently.
2. The dimensions of interest for each mating component follow a
    normal distribution (the manufacting process was under con-
    trol, i.e., consistently meeting the specified tolerances.)

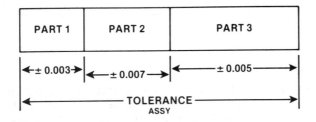

(a) **STRAIGHT ADDITION:**

$$T_A = T_1 + T_2 + T_3$$

$$T_A = 0.006 + 0.014 + 0.010$$

$$T_A = 0.030 \ (\text{MAX./MIN.})$$

(b) **SUM OF SQUARES**

$$T_A = \sqrt{(T_1)^2 + (T_2)^2 + (T_3)^2}$$

$$T_A = \sqrt{(0.006)^2 + (0.014)^2 + (0.010)^2}$$

$$T_A = \sqrt{0.000332}$$

$$T_A = \quad 0.018 \ (\text{STATISTICAL})$$

FIGURE 10.8 Simple calculation of statistical tolerancing.

3. The parts are selected randomly for assembly.
4. The actual dimensions are approximately centered around the nominal specification values.

This method can be used in both directions. It can show what the assembly tolerance would be for given component tolerances. Or it can be used to explore what combination of component tolerances could be most cost-effective for achieving a desired assembly tolerance. Since it may be more economical to manufacture a tight tolerance on one type of component than on another part in the assembly, different tolerances can be applied to the various components in the assembly and still achieve the desired assembly tolerance range.

10.6 SUMMARY

Since variation is present in all things, it is important for the designer to examine the effects of variation on product design. Several simple, but powerful, tools are available for assessing the effects of variability.

This chapter highlights the use of frequency distributions, the calculation of averages and standard deviations, process capability studies and statistical tolerancing in design and development activities. Each of these tools provide insights in the degree of variability present and its impact. With this information, the designer can better appreciate whether or not the design requirements are being satisfied.

Several good texts on statistics are listed in the section on Selected Readings and are recommended as detailed sources of information on these and other statistical tools, such as probability studies, hypothesis testing, design of experiments and analysis of variances.

11

Control of Nonconformances

11.1 Principles of Control   198

11.2 Nonconformance Reporting   201

11.3 Disposition of Nonconformances   203

11.4 Closeout and Feedback   204

11.5 Summary   205

11.1 PRINCIPLES OF CONTROL

Up to now, the emphasis has been on doing the work right. However, occasionally things do go astray after the design has been released for production. Some requirements may not be met, but all such nonconformances need to be controlled. In this chapter you will see that the control of nonconformances is similar in many ways to the control of changes, which were discussed in Chapter 5.

Nonconformances can occur in design, manufacturing or procurement. They may be found in the materials, parts or completed units and can affect the form, fit or functioning of the item. Each nonconformance represents a potential problem to the user. As such, the designer must be concerned with both the cause and the solution.

In this chapter, the term "nonconformance" will be used to describe all conditions which do not meet the drawing and specification requirements. Although other terms, such as discrepancy, deviation, deficiency, defect or error, may be used in industry, the author believes "noncomformance" is more appropriate for general use and does not imply something necessarily ominous about the condition. In many instances, the nonconformance may be quite harmless and inconsequential.

198

However, in other cases the nonconformance may have a significant effect. Thus, each nonconformance must be investigated as part of the control process.

There are several basic principles that must be included in the process for controlling nonconformances. These are described below.

Identification: When a nonconformance is discovered, the items affected must be clearly identified as being in noncompliance with a specific requirement. This is usually accomplished by attaching a special tag to the item or marking the item to show it is nonconforming.

Criticality: Each nonconformance must be treated as if it will have serious consequences until dispositioned otherwise, (i.e., guilty until determined it is innocent).

Segregation: All nonconforming items should be physically separated from conforming items, wherever possible. If it is not possible to segregate due to size or space limitations, it is imperative for the nonconforming items to be clearly marked. This is to reduce the chances that the nonconforming items would be inadvertently used without proper authorization.

Documentation: A written description of the nonconformance should be made for each instance of noncompliance with the drawing and specification requirements.

Reporting: Each nonconformance must be reported to the management of the organization where the nonconformance was discovered (whether or not that group caused the nonconformance) and to the engineering organization who is responsible for deciding what actions must be taken. The Quality Assurance organization should also be notified if they were not the ones to identify the nonconformance initially. Notification is frequently accomplished by distributing a copy of the nonconformance report to the different groups.

Investigation: The responsible engineer, normally the designer, must investigate the impact of the nonconformance. This includes determining the effect on the final product's ability to perform its intended function properly and safely, as well as the effect on subsequent manufacturing operations.

Disposition: After investigation, the responsible engineer then decides what actions, if any, are required to resolve the problem. Table 11.1 lists the typical dispositions and their interpretation. Although other groups, such as quality assurance, manufacturing, production control, purchasing, etc., may be involved in the overall disposition process, the responsible engineer must be charged with the final authority to use any nonconforming items. Others may decide to discard, but only the designer (or the person designated by the engineering management) should be permitted to authorize the use or modification of nonconforming items.

In large organizations, the specific assignments of engineers change frequently, and it may not be possible to maintain the same

TABLE 11.1 Typical Dispositions of Nonconforming Items

Disposition	Interpretation
Use as is	Use the nonconforming material in its existing condition without taking any further action.
Rework	Take action to bring the item into full compliance with the applicable drawings and specifications (e.g., drill the hole that was omitted).
Repair	Take action which will make the item fit for its intended use, although it will still not conform exactly to the original drawing and specification requirements (e.g., weld up the hole that was drilled in the wrong place and redrill the hole in the right place. Drawing does not call for a welded-up hole).
Scrap	Discard the nonconforming items, since it is not possible or not economically feasible to salvage (e.g., throw the part away and obtain a new one).

engineer in a project assignment for the life of the project or product. However, it still is necessary that the assigned engineer be competent in the area of responsibility, have an understanding of the applicable requirements and have access to the pertinent files and information on the subject. The responsibility for making a proper disposition must not be taken lightly or treated casually.

Closeout: Each nonconformance must be closed out by the following steps: (1) take the necessary actions in accordance with the engineer's disposition; (2) for items which are reworked or repaired, verify by inspection or test that the items in question now conform to the engineer's disposition; (3) for items which are now acceptable, remove the nonconformance tag or marking and release for use; (4) for items which cannot be made acceptable, see that these items are discarded and are not inadvertently (or intentionally) used; (5) mark the records to show how the nonconformance was resolved and when.

Administrative Control: The specific methods for controlling the identification and resolution of nonconformance must be defined in a written procedure for all involved groups to use and follow (manufacturing, engineering, quality, production control, etc.).

Consistent application of these principles will assure proper control of nonconformances. Anything less may result in a breach in product integrity.

11.2 NONCONFORMANCE REPORTING

Various types of nonconformance reports are used in industry today, although two general types are most common. The first of these is a multi-sheet card as shown in Figure 11.1. This type of nonconformance report consists of one or more carbonless paper sheets with the back sheet being made from card stock. The information is handwritten on the top sheet and reproduced on the other sheets and the card. The paper sheets are then torn off and distributed to the specified groups, and the card is physically attached to the nonconforming items. The card remains attached until final disposition is made and verified.

The nonconformance tag is typically a one-way document. It simply identifies and describes the nonconformance condition. A separate document (sometimes referred to as a waiver) is often required for dispositioning the nonconforming items. This system is commonly used when the nonconforming items are physically located within a factory or manufacturing facility, and the responsible engineering group is close by.

The second type of report is a more conventional form. It may be a single sheet or multi-part snapout form, frequently 8-1/2 x 11 inches

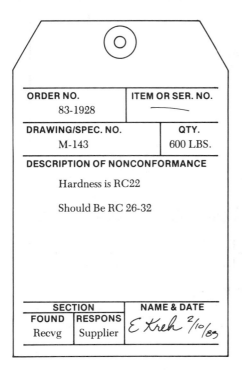

FIGURE 11.1 Nonconformance tag.

ASTON MANUFACTURING CO. ASTON, NY		NCR NO. 1643	
PART NO. 909E641	**PART NAME** Vane Shroud		
PROJECT XJ215	**ORDER NO.** 79-20419	**QTY. AFFECTED** 14	

DESCRIPTION OF NONCONFORMANCE & PROBABLE CAUSE

Radius is 12.630 - 12.645 In.

Should be 12.660 - 12.700 In.

SOURCE OR SUPPLIER AJax Mfg.	**REPORTED BY:** *Bill Lewis* 8/9/83

DISPOSITION

Will not fit.

Return to supplier.

APPROVED BY: *L Weyley* 8. 10. 83	**APPROVED BY:** *Ken Edwards* 8/11/83	**APPROVED BY:** *F Theodore* 8/12/83
ACTION TAKEN Returned to Supplier	**VERIFIED BY:** *Jim Kester* 8-17.83	

FIGURE 11.2 Nonconformance report.

in size. Figure 11.2 illustrates a sample of this type of nonconformance report. It can be used as the waiver form described above, or it can be used in situations where the nonconforming items are found in a location geographically separated from the responsible engineering organization. For example, a supplier may discover a nonconformance which affects the buyer and submits such a report to request disposition. The form normally contains space for writing disposition instructions and serves as a two-way document. Another frequent use of this type of document is for a customer to report back to the supplier that a nonconformance was discovered after delivery and corrective action is required. In this situation, the supplier will normally be requested to define what will be done to correct the current problem and to prevent recurrence of this same condition in the future.

The prime thrust for design assurance purposes is to require that all nonconformances with drawing and specification requirements be reported promptly in writing.

11.3 DISPOSITION OF NONCONFORMANCES

As described in previous sections, it is absolutely imperative that the cognizant engineer be the person designated to make the disposition decision. However, this can be accomplished several ways. In small organizations, the engineer simply goes to the manufacturing area or receiving area where the nonconformance was discovered, discusses the problem with the shop personnel, examines the items, determines the effect, and makes the disposition.

For more complex or costly items, representatives from other departments may have to provide information about the problem and its impact on production (e.g., inspection, production control, purchasing, etc.) and still others to assist the engineer in evaluating the technical effects of the nonconformance on the product and its application (e.g., materials specialists, structural analysts, welding engineers, etc.).

One frequently used method is the Material Review Board (MRB). The MRB normally consists of a representative from production, quality assurance and engineering. In large plants or projects, the MRB may meet daily or several times a week to review and make decisions on the various nonconforming items. Where the number of nonconformances is less, or where specialty assignments are required, the MRB may meet on an as-needed basis.

The members of the MRB investigate the problem, its cause and its effects. The Engineering representative evaluates the various inputs and makes the technical decision. The full MRB membership then determines the course of action required to implement the engineer's technical decision, e.g., define the repair process, order new parts, release the parts to be used "as-is", etc.

In some instances, the nature of the product may be such that the customer insists on participation in the disposition of certain types of nonconformances. These typically involve characteristics defined in the customer's drawings or specifications, such as performance or operating characteristics, dimensions which affect mating parts, items which may degrade safety or reliability, etc. Such nonconformances have a similar impact as described for a Class I design change in Chapter 5.

When the customer has specified this right, the supplier must make provisions to accommodate this. In reality, suppliers tend to discard all but the least significant nonconforming items, since customers typically are not interested in accepting nonconformances, affecting contractual requirements.

11.4 CLOSEOUT AND FEEDBACK

After disposition of the nonconformance is made, all records of the open item must be closed out. This may be a simple listing maintained by manufacturing, quality assurance or engineering; or it may require specific actions to enter the information in the computer data system; or it may require a written notice be sent to the customer or supplier.

Except for items dispositioned "use as is" or "scrap", two types of corrective action are normally needed. The first type is the short-range action to fix the items found to be nonconforming. This involves some form of rework or repair, plus a followup inspection or test process to verify the items now comply with the disposition instructions.

The second type of corrective action should be directed at identifying the root cause of the problem and implementing measures to prevent the condition from recurring in the future. Such causes may stem from how the requirements are specified, the condition of the material or parts received from the supplier, how the work was performed in the factory, a worker's error, a broken or incorrect tool, etc. But whatever the cause, the engineers (and local management) should be interested in seeing that the true cause is identified and action taken to correct the problem once and for all. It may require actions by Manufacturing, Purchasing or Engineering to eliminate root causes of problems, but such effort is a cornerstone in all defect prevention programs.

However, for Engineering to take appropriate action, they need feedback about the nonconformances. This information is usually compiled and issued by the quality assurance group. The data should be issued two to four times a year, depending upon activity level, and must be factual. The reports need to define the drawing or specification number affected, a summary description of the nonconformance, and the disposition. From this, the engineers can look for patterns of recurring nonconformances which should be investigated.

One type of recurring problem is where the requirement is incorrect or sufficiently unclear that other departments are taking wrong or improper actions. Problems such as this require an engineering change to revise the drawing or specification, and these should be processed promptly.

The second type of engineering corrective action is related to design assurance but more directly involves good engineering practices to reduce costs. The Engineering Department should periodically review the nonconformance summary reports to identify which items were frequently dispositioned "use as is". Recurring cases where the same requirement is not met, but the item is still considered usable, should be questioned. It may be the drawing or specification requirement is unnecessarily tight and could be relaxed. If so, this can avoid the recurring costs of processing the nonconformance reports. It also eliminates the associated delays and interruptions in production and the waste of people's time associated with identification, marking, segregating, reporting and dispositioning the nonconformance. All of this represents unproductive effort and should be avoided to the maximum extent possible.

Some companies use the rule of thumb that after three acceptance decisions for the same condition the drawing must be changed. Unfortunately, some engineers don't want to be bothered by correcting these kinds of problems. However, in the long run, efforts taken by Engineering to revise drawings and specifications for the purpose of preventing recurring nonconformance contributes to both improved design integrity and cost reduction. It is not something to be set aside and forgotten. If some specified requirements are continually missed but accepted, there is a tendency for people to think other requirements fall in this same category. In the long-run, it can lead to a degrading of the product. The guiding principle should be: rigorous enforcement of reasonable requirements.

11.5 SUMMARY

Proper control of nonconformances is another important element in an effective design assurance program. Design engineers play a key role in the evaluation and decision of what to do with nonconforming parts or materials. Their technical knowledge and judgment must be used in determining whether or not the nonconforming items can be used or modified or must be discarded. No one else is really qualified to make those decisions.

Controlling nonconformances requires using specific procedures for identifying, reporting and dispositioning items which do not comply with the applicable drawing and specification requirements. It also requires the organization to follow these procedures closely to maintain product integrity. Further, the process obligates Engineering to correct any of the design requirements which cause or contribute to the occurrence of the nonconformance.

All of these actions are needed to assure the final product will fulfill the design requirements.

12

Engineering Records

12.1	Introduction	207
12.2	Methods for Indexing	208
	12.2.1 Alphabetical Indexing	208
	12.2.2 Numerical Indexing	210
12.3	Filing Practices	211
12.4	Records Retention	218
12.5	Summary	220

12.1 INTRODUCTION

Records play an important role in every engineering department. They document what has been done in the design process and describe the product in the form of drawings and specifications. Many other documents, such as data forms, calculation sheets, test reports, field performance data, etc., are included in the engineering files to support the design process. Even though these records are the building blocks of the department's work, the process for organizing and maintaining the records is frequently neglected.

Some firms set up central files for their records, while others depend upon each engineer to maintain his own files. Regardless of the approach followed, certain methods and practices can greatly aid the process.

An effective records system reduces the chances of important documents being lost and helps in finding needed information in the files. A set of basic rules and decisions are needed, but they are reasonably straightforward and easy to follow.

The main questions to consider are: What records should be kept? Who keeps them? How to catalog and file? And how long to keep them?

For a small engineering department, the records system can be simple, but a large department with many sections or projects may require a more elaborate program. Nevertheless, the process needs to be pre-planned, systematic and readily understandable. The basic principles are the same for both large and small records programs, and these can be applied by the individual engineer or a central records group. This chapter presents guidelines and preferred practices for the basic elements of an engineering records program, i.e., indexing, filing, storing and retrieving.

12.2 METHODS FOR INDEXING

There are two basic systems for indexing—alphabetical and numerical. Each is adaptable to most circumstances, although each method is better suited for certain types of records. Both methods are described in this section, and the pros and cons for each are presented.

12.2.1 Alphabetical Indexing

This method uses the letters in names or topic words for indexing purposes. Customer's names, subject headings, or geographical locations are examples of information which is often filed in alphabetical sequence.

Two different styles can be used for alphabetical filing. The first is referred to as the dictionary style. It requires that each item be filed in absolute alphabet sequence. An example of this arrangement is shown in Figure 12.1(a).

The other frequently used type of alphabetic filing is the encyclopedia style. This groups related items under a series of general headings. Both the primary and secondary headings are then arranged in alphabetical sequence. For example, various manufacturers' catalogs could be grouped by like products: bearings, castings, forgings, motors, etc. Figure 12.1(b) illustrates how the same data shown in Figure 12.1(a) would be arranged in the encyclopedia style.

Alphabetical filing has several advantages. These are: (1) the ability to expand indefinitely; (2) the process is self-indexing; and (3) the ease of classifying material for filing, especially names. The major limitation of the alphabet system is the need to define and follow specific rules for alphabetizing. Several texts on filing are available and define the rules and practices in detail. Rules for the most commonly occurring situations are summarized in Table 12.1.

Another problem is sometimes encountered when the encyclopedia method is used. There seems to be a tendency for overlapping file categories to creep into the system under two or more topic headings. For example, one document on steel castings might be filed in a folder for "Castings" under the general topic of "Manufacturing", whereas a similar document might be filed in a folder for "Steel Castings" under

(a) DICTIONARY STYLE	(b) ENCYCLOPEDIA STYLE
A	*Administrative*
Aluminum data	Budgets
	Charge numbers
B	*Design data*
Beams	Beams
Blowers	Columns
Budgets	
C	*Equipment*
Charge numbers	Blowers
Columns	Compressors
Compressors	Dynamometers
D	*Materials*
Dynamometers	Aluminum data

FIGURE 12.1 Example of dictionary and encyclopedia styles of indexing engineering records.

TABLE 12.1 Rules for Alphabetizing

File each name in exact letter sequence.

Spell out abbreviations
   "ASME" filed as "American Society of Mechanical Engineers"

Arrange names of persons and companies with last name first
   "A. B. Cline and Co." filed as "Cline, A. B. and Co."

File trade names in original word sequence.

File names beginning with M', Mac or Mc in exact letter sequence.

File documents from government agencies first by name of government (United States, State of Iowa, etc.) and then by name of agency or department.

Nothing comes before something
   "Engines" filed before "Engines, Diesel"

NOTE: The telephone directory is a handy reference for alphabetizing practices.

From John A. Burgess, File Now-Find Later, *Machine Design*, Penton/ IPC, Cleveland, Ohio, April 28, 1966.

the general topic of "Materials". Thus, it is necessary to use care in
developing and maintaining exclusive file categories in the index.
Filing must then be done according to these rules to prevent the over-
lap from developing.

The encyclopedia style is generally well-suited to engineering subject files. It simply requires some discipline in developing the system
and using it consistently. More on this is presented in Section 12.3.

12.2.2 Numerical Indexing

Certain kinds of information are difficult to file alphabetically. For instance, whenever there are many similar, but unrelated documents, or
uniquely serialized documents such as drawings, specifications, reports, purchase orders, etc., the alphabetical type system is clumsy
to use. For such applications, a numerical sequence file is the obvious
choice. In this system, each similar document, e.g., drawings, is
filed in its assigned number sequence.

The sequential numbering scheme is the easiest of all indexing
systems to use for filing purposes. Also it is well-suited for retreiv-
ing a specific document when the proper number is known. However,
some form of cross-index between numbered items and subject or application may be needed in the event the specific document number is
not known. The sequential numbering system has one other advantage.
It shows if there are voids or missing documents within the number
sequence.

Some engineering groups seem to favor assigning numbers to their
subject or topic files. These systems tend to be like a Dewey decimal
system but with locally designated topic categories. Those advocating
this approach claim it uniquely identifies each file and eliminates the
indecision about alphabetizing. However, such a system requires that
a separate index, showing the number of each file topic, be maintained
and used. These systems tend to become outdated with time as new
files are needed and created. It is often difficult to add new files without causing topic overlaps with existing files.

Numerical systems have one other hazard. There seems to be a
natural tendency for people using numerical subject files to attempt to
memorize the index numbering system. Rather than looking up the
file number in the index when classifying a document for filing, they
often rely on their memory for the file number. This frequently results in the assigning of an incorrect file number, leading to a misfile.
When a misfile occurs in a numerical subject file, it may be difficult to
locate, since the misfiled document may end up in a totally unrelated
file and, due to the numbering system, it may not be in close proximity
to the correct file. In contrast, a spelling error or similar misfiling of
alphabetized documents tends to be more localized, making it easier to
find a misfile.

Files with consecutively numbered documents also have one other problem; related but distinctly different types of documents, such as drawings and correspondence for a given project, are not grouped for convenient use.

Obviously, no one system easily handles all situations. Therefore, it is necessary to use combinations of indices and filing practices to suit your particular needs. Suggestions and recommendations for indexing and filing various categories of engineering documents are presented in the next section.

12.3 FILING PRACTICES

Although the volume and exact detail of engineering records may vary widely among companies, there is a lot of commonality in the general types of records used by many engineering organizations. Consequently, some general guidelines can be developed and applied for most firms. Table 12.2 presents Do's and Don'ts of good filing practices that fit most situations. In addition to these general rules, Table 12.3 presents recommendations for indexing and filing the various types of records frequently found in engineering departments. These recommendations are not meant to be the only way the documents can be filed, but the methods presented have been found workable in numerous applications in industry. They should at least be considered when developing a new filing system for your engineering organization.

Engineering reference files present a special challenge, especially when it must serve several or many different engineers. Not everyone thinks alike or applies the same logic when considering where to put a particular document for future reference. This is where some thought and pre-planning is needed.

The key ingredient is a clear and specific index for all to see and use. Table 12.4 lists the series of steps to follow when developing such an index. Once completed, the index must be modified each time there are additions or changes to the filing categories. Otherwise, the ability to find documents in the future will degrade quickly. Figure 12.2 provides one example of how such a subject file index might look.

Engineering groups that work on many similar but different projects often use a specialized type of file. Each job is assigned a unique project or job number. This number then becomes the key to a "one-spot" filing system. All records associated with that job are collected in a pre-planned set of categories and are maintained in one set of files. Table 12.5 provides a typical list of information that might apply to each job. As documents are received or created for a given project, the job number and file category are assigned to each document. The documents are then filed in the pre-designated set of folders for that project. This approach uses a significant-number system similar to that described in Chapter 3 for projectized drawing number systems.

TABLE 12.2 Do's and Don'ts for Better Filing

Do:	use plenty of dividers and file behind the dividers.	Don't:	file more than ten folders per divider.
Do:	stagger divider tabs for easy viewing.	Don't:	use colored tabs unless needed for visibility and ease of access.
Do:	keep papers neatly in folders. Line up papers along top and left margins. Always put letterhead on left side of drawers.	Don't:	overstuff folders. Limit to approximately one-half inch of paper.
Do:	allow for expansion.	Don't:	let the index become obsolete.
Do:	use staples to fasten related papers together.	Don't:	use paper clips. They tend to pick up unwanted papers.
Do:	use drop filing with latest document on top for speed filing.	Don't:	fasten papers in folders, except for very important documents.
Do:	use OUT cards for charging out documents removed from the file.	Don't:	let inexperienced personnel replace documents in files unless closely supervised.

From John A. Burgess, File Now-Find Later, *Machine Design*, Penton/IPC, Cleveland, Ohio, April 28, 1966.

TABLE 12.3 Recommended Indexing/Filing Practices for Engineering
Records

Drawings: File by drawing number sequence

Numbers only	Letters, then numbers
517832	963B403
517835	964B122
593114	832C719
600771	435D345

Specifications: File by specification number sequence for similar
types (materials, process)

M-1819	Steel, carbon
M-1823	Steel, corrosion resistant
M-1907	Aluminum, age hardening
P-1043	Plating, hard chrome
P-1115	Welding, submerged arc
P-1249	Plating, zinc

Cross-reference specifications by listing on 3 x 5 inch file cards alphabetically using main noun in title:

Aluminum	M-1907
Steel, carbon	M-1819
Steel, corrosion resistant	M-1823
Plating, hard chrome	P-1043
Plating, zinc	P-1249
Welding, submerged arc	P-1115

Reports: File in report number sequence

81-473	Mechanical design of model X-12 gearbox
81-492	Design studies of hydrostatic drives
81-503	Failure mode analysis of model 24 transmission
82-176	Test report: model X-12 gearbox

Cross reference either by filing a copy of the title/abstract page in the
subject reference file, or by maintaining a card file with report titles
filed alphabetically. Use main noun or noun phrase in title for filing
and list report number.

Gearbox, Model X-12, Mechanical Design, Report No. 81-473
Gearbox, Model X-12 Test Report, Report No. 82-176
Hydrostatic Drives, Design Studies of, Report No. 81-492

Patent Records: (Patents/Disclosure files) File patents and related
documents in numerical sequence by U.S. Patent

TABLE 12.3 (Continued)

| | Office number. File disclosure records in numerical sequence by company-assigned disclosure number. |

Trade Catalogs: File alphabetically by subject and manufacturers' names.

Bearings	Fafnir
	SKF Industries
	Torrington

Gears	Boston Gear
	Philadelphia Gear
	Sier-Bath

Can also use a card file for cross referencing trade names or manufacturers who describe several different products in the same trade catalog.

Technical Clippings and Reprints: File alphabetically by subject. Mount clippings on 8-1/2 x 11 sheets before filing.

Engineering Change Notices: File in numerical sequence. If large volume, file together. If small volume, file in applicable project or product folder.

Project Records: File in numerical sequence by job number or alphabetical by project name.

Correspondence: Several different methods can be used for incoming and outgoing letters and memoranda.

File by subject with latest date on top.

File by job number with latest date on top.

File both incoming and outgoing correspondence together in a "Correspondence" folder chronologically with latest date on top.

Calculations: File by project number or alphabetically by main subject and subtopic.

Frame	Column deflection
Spindle	Critical speed limits
Spindle	Torsional frequency

File chronologically if books or voluminous computer printouts are used.

Can also file in numerical sequence when specific calculation numbers are used. Good idea to cross index by card file or computer listing

TABLE 12.3 (Continued)

Purchase records: File in numerical sequence by purchase order number or compile by job number (latest on top).

Discrepancy records: For standard product lines, file by part name or part number with latest document on top.

For specialty products or job shop work, file by project with latest document on top.

Procedures: File by procedure number or by category of procedure.

Budget/Cost records: File in chronological sequence by department or project.

Personnel records: File alphabetically by employees' last names.

From John A. Burgess, File Now-Find Later, *Machine Design*, Penton/IPC, Cleveland, Ohio, April 28, 1966.

TABLE 12.4 Guidelines for Developing a Subject File Index

---

1. Survey the various items which are representative of the activity. List everything that comes to mind as you think of it.

2. Take the total list and group related items.

3. Review the groupings and assign each a suitable title. Choose titles carefully to prevent overlaps in file categories. Use two or three-word descriptive noun phrases.

4. Arrange the titles in alphabetical sequence.

5. Arrange the items under each title heading in alphabetical sequence. This is now the basic file index.

6. Prepare a set of dividers for each category heading.

7. Prepare a set of file folds for each subheading. Also prepare a "General" folder for each category heading to handle the odds and ends that do not warrant a separate specific folder.

8. Classify each document according to the index and file it in the corresponding folder.

From John A. Burgess File Now-Find Later, *Machine Design*, Penton/IPC, Cleveland, Ohio, April 28, 1966.

INDEX ENGINEERING DEPARTMENT FILES

File Category	Content guidelines	File folder title
Associations	Technical societies and trade associations. Includes standards and technical committee activities.	ASME - General ASME - Compressor technical committee IEEE - General Metro Management Club NEMA general
Committees	In-company committees and task groups. Includes meeting minutes, plans, reports.	Plant expansion task team Product planning committee Safety committee
Customer projects	Summary data for each project. General plans, schedules, cost data, status reports.	Alexander Bank Atlas Manufacturing Co. Benson Bakery Goldstein, Inc.
Energy resources	Contains data and information on energy costs, requirements, special provisions.	Electric power Natural gas
Equipment	Includes general and technical data for components and systems used in company's projects.	Belts/pulleys Compressors Condensers Controls, electrical Controls, pneumatic Dampers Expansion valves

FIGURE 12.2 A sample descriptive subject file index.

The reader is also referred to Chapter 6 for the method of filing calculation sheets by serialized document number and cross-indexing through a computerized, key word system. This concept can be used for other documents generated in large quantities but requiring periodic access and retrieval of individual documents.

From a design assurance standpoint, the primary interest in the records system is to see that important records which support the design are properly maintained. It is equally important to be able to find and retrieve these records whenever they are needed in the future. Thus, even though filing may seem like a trivial detail, outside the scope of engineering, it is not something to be forgotten or ignored. It only takes one serious customer complaint or field problem that could be resolved if only you could find the old files to make a believer out of a skeptic. Indexing and filing practices frequently make the difference. How do yours measure up?

TABLE 12.5 Typical Project File Categories

Category	Contents
1. Requirements:	Contracts, statement of work, special instructions.
2. Costs/schedules:	Cost estimates, budgets, financial reports, time estimates, project schedules.
3. Correspondence:	Incoming and outgoing letters and memos, arranged in chronological sequence with latest dated item on top.
4. Technical data:	Calculations, suppliers' spec sheets, test data, design studies, performance data.
5. Drawings/specifications:	Design documents created for or applicable to accomplishing the project. Includes design change orders.
6. Procurement documents:	Requests for quotations, supplier quotes, purchase orders, work orders, procurement change notices.
7. Quality documents:	Inspection records, material certifications, discrepancy reports.
8. Project summary:	Final reports, parts lists, photographs, illustrations, instruction manuals.

From John A. Burgess, File Now-Find Later, *Machine Design*, Penton/IPC, Cleveland, Ohio, April 28, 1966.

12.4 RECORDS RETENTION

As the heading implies, the engineering records system should also address what items are to be kept, where and for what period. The process is best handled in a two-step approach. The records are divided into two categories of active and inactive records. Active records are typically the most recent documents and are frequently used in the day-to-day conduct of business. As such, these records should be kept in a location which is convenient for access and use of the documents.

The inactive category applies to those documents which have long-range intrinsic value to the department or the project but are no longer needed for frequent or immediate reference. These documents can be moved to another location for long-term retention.

In small engineering organizations, engineers frequently keep the active job files in their desks. It is convenient for them, but it would be better to organize and store the folders in a designated file cabinet nearby. This provides access for others that may need to refer to the information and better physical protection of the records. Large engineering departments frequently use central files for nearly all of their engineering records. This is necessary for accurate filing, ready retrieval for persons needing access to the information, and greater control and protection of the records.

To the maximum extent possible, engineers in both large and small departments should be discouraged from keeping active or inactive job files in their desks.

Just how long should engineering records be kept? It is a key question, but it is difficult to answer for each individual document. However, general guidelines for records retention times have been developed from experience for typical Engineering Department records. These are presented in Table 12.6. Since the laws governing statutes of limitations vary from state to state, and may have a significant effect on your records retention requirements, consult your legal representative when finalizing local retention periods for specific types of documents you use.

There are four key steps to follow when developing a records retention plan. First, determine the various types or categories of records in the files and compile these into a list. Next, establish specific active and total retention periods for each record category, such as shown in Table 12.6. Then define the method and process for transferring records at the end of their active period to inactive storage. This includes the physical movement of the records from one location to the other. Finally, establish a plan for the destruction of obsolete files at the end of the designated total retention period. Provisions should be made to require a review of files which have reached the end of the inactive storage period, prior to actually discarding or destroying the records. There is a possibility that the circumstances

TABLE 12.6 Representative Retention Times for Engineering Records

Document	Active period (Years)	Total period (Years)
Drawings	5	Permanent
Specifications	5-10	Permanent
Correspondence	1-3	5-10
Completed project summary folders	1-3	Permanent
Proposals	1	5-10
Purchase orders	1-2	3-6
Engineering reports (Bound)	3-5	Permanent
Design change orders	1-2	5-10
Design calculations	1-3	Permanent
Patent records	1-3	Permanent
Parts lists	5	Permanent
Budget/Cost data	1-2	5
Procedures	As long as valid	5 (after active period)
Personnel records	During continuous service	Permanent

From John A. Burgess, File Now-Find Later, *Machine Design*, Penton/ IPC, Cleveland, Ohio, April 28, 1966.

may have changed after the retention period was originally specified. Thus, it may be necessary to keep certain records longer than originally planned. Marketing and legal representatives should participate in the review to advise if there are any special requirements for further retention of the records. If there isn't, instructions should then be given to discard the documents (usually by shedding or burning).

Recognize that long-term storage and purging of records is both a strategic and economic decision. It costs money to store records. On the other hand, the records may be valuable to the business to support the on-going design effort and to fend off future claims against the product design. Each company must weigh these costs and risks and make their own decisions accordingly.

One possible approach is to use some form of microfilming or computer storage of the data to reduce volume and space requirements but maintain the records for strategic purposes. These methods also have costs associated with them, so the cost-benefits of each approach also should be considered.

One other factor should be recognized in your records retention planning. That is physical security. What degree of protection is needed against accidental or catastrophic loss, such as water damage, fire, theft, mice, etc.? Appropriate measures must then be provided. These might consist of physical separation of duplicate records, use of safe deposit boxes or similar protective actions. Again, security provisions must be considered on a cost-risk basis, comparing the cost of protection vs. the potential consequences of loss.

The most difficult decision in any records program is determining what records and documents need to be saved. This can only be answered individually by each engineering department. The guiding theme is to retain those records which define your product and those that support the decisions made during the design and evaluation of it. These documents then provide the evidence of quality that went into the design. All such records are candidates for storage and protection.

12.5 SUMMARY

Maintaining records is frequently regarded as the mundane side of engineering. No one wants the job or typically pays much attention to it. Yet records are vital for supporting the design. Key engineering documents demonstrate the adequacy of the design and development process.

Each engineering organization can benefit by using an orderly and systematic method for indexing, filing and retaining their engineering documents.

It is not a frill but a necessity for most engineering departments today. You might ponder these questions: Do our files adequately support our design efforts? Can we find the data when we need it? Is it adequately protected against loss? If the answer is "no" to any of these questions, it is time to take action to improve your engineering records system.

13

Supporting Documentation

13.1 Introduction   221

13.2 Operation and Maintenance Instructions   222

    13.2.1 Types of Instruction Documents   222
    13.2.2 Typical Content   223
    13.2.3 Guidelines for Writing   225

13.3 Replacement Parts Lists   228

13.4 Design Control Measures   231

13.5 Summary   232

13.1 INTRODUCTION

Installation, operation and maintenance may seem far removed from the design of the product. However, improper use can negate much of the hard work that went into the design. Unfortunately, customer misuse of the product is often due to the manufacturer failing to instruct or caution the user adequately. Or, the manufacturer's information may be erroneous, incomplete or misleading.

It is this aspect that brings the user documentation into the design assurance arena. The supporting documentation given to customers and users must be as carefully engineered and controlled as the product itself. Otherwise, shortcomings in installation, operating or maintenance instructions can cause both the user and manufacturer a lot of grief.

For design assurance purposes, the supporting documentation needs to be addressed in a manner similar to the basic design drawings and specifications. Items, such as instruction manuals, replacement parts lists, and related technical literature should be developed from a

clear set of requirements, checked carefully for completeness and accuracy, and subjected to a level of review and approval consistent
with that applied to the product. In addition, changes must be made
in a timely manner to maintain compatibility with the product design
and its subsequent revisions.

This chapter presents basic information for the preparation and
control of the supporting documentation for an engineered product.
It identifies key ingredients and methods for assuring that the quality
of design has a reasonable chance of being maintained after the product is in the customer's hands.

13.2 OPERATION AND MAINTENANCE INSTRUCTIONS

One of the most important pieces of information to the customer is how
to use and service the equipment. The significance of this data increases with the complexity of the product. Department of Defense
studies from the 1960s and 1970s show that cost of operation and maintenance over the life of the product greatly exceeds the original cost.
Sometimes, it was as much as 500% of the purchase price. Thus, the
users of complex equipment want to keep down the cost of ownership.
They want to know how to operate and maintain the equipment properly
and minimize the expense, delay and loss of use associated with servicing and repairing it.

In many industries the manufacturers provide very little information about the equipment they sell to their customers. As a result,
many errors and mistakes are made which affect both the user and the
manufacturer. Incomplete, inadequate or missing instructions can
cause the user to operate the equipment incorrectly, perform improper
or incomplete maintenance, order the wrong replacement parts, lose
the use of the equipment, and tie up maintenance personnel and facilities to keep the equipment operating. This can also lead to excessive
warranty claims, field service expenses and damage to the manufacturer's reputation.

Many of these problems can be avoided by developing and providing meaningful supporting documentation for the users. The various
types and typical content of instruction manuals are described in the
next section.

13.2.1 Types of Instruction Documents

There are several different types of instructions used to support major
systems or complex equipment. These include:

Installation Instructions: Describe how the equipment should be
installed for use. This includes directions for unpacking the items,
making the necessary fitups and connections and attaching it to the
necessary supports, foundations or mounting points. It may also

provide instructions to make an initial checkout of the equipment to see if it will operate properly.

Operation Instructions: Define how to operate the equipment over the normal range of use. For complicated equipment, this may require very detailed, step-by-step instructions with appropriate cautions and warnings noted. Some equipment may have several modes of operation e.g., startup, standby, normal use, emergency operation, etc. In this event, instructions must be given for each mode of operation.

Calibration Instructions: Explain how the equipment must be checked and adjusted to achieve optimum operation within the inherent accuracy and precision of the equipment design. These instructions should also define what other equipment must be used to perform the calibration operations, e.g., meters, loading sources, connections, etc.

Servicing Instructions: Describe how to perform those operations needed to keep the equipment operating normally in its useful lifetime. This includes fueling, lubricating, adjusting clearances, aligning parts subject to wear or multiple settings, replacing worn or limited-life parts, such as belts, fuses, etc. These tasks are normally restricted to actions the user is expected to perform. The tasks typically can be accomplished by the average person without needing any special tools.

Maintenance Instructions: Define how the equipment can be returned to its original state of operation after certain types of failure, breakage or wearout conditions occur. Depending upon the nature and complexity of the equipment, the maintenance instructions generally require levels of skill and tools which are beyond those possessed by most users. However, even maintenance instructions now tend to be limited to those levels of repair which can be accomplished by modular replacement of detachable components or subassemblies. Complete teardown or disassembly to the detailed piece part level is usually performed at a depot or specialty facility. Such actions typically require even more detailed instructions.

Repair/Overhaul Instructions: Describe how to disassemble and rebuild the equipment in a manner comparable with those operations performed by the original equipment manufacturer. The repair/overhaul instructions routinely are very detailed and specify the use of specialized tools, equipment and machines. The work typically must be done by specially trained mechanics or technicians.

For simpler equipment, it is common to combine several of the different categories of instructions into one document. In some cases, only very simple instructions are needed at all. These may be limited to information printed on the package or described by a set of instructions packed with the equipment.

13.2.1 Typical Content

For many engineered products, the user instructions are contained in a single document. Such a document may be called Operating and

Maintenance Instructions, Instruction Book, Owner's Manual or similar title. The document generally will be prepared in booklet form and is intended to be retained by the user for the life of the product.

Listed below are the recommended contents of a typical user instruction booklet.

Introduction: Explain the scope and purpose of the booklet. It describes the equipment and defines the various models or styles covered. It is a good practice to include photographs in this section of the actual equipment for clarity and identification purposes.

Theory of Operation: Define the basic principles of how the equipment works. The Theory of Operation section should explain the logic or information flow or present the sequence of events as the equipment operates through its duty cycle (e.g., startup, standby, run, shutdown). Since this section is often used for troubleshooting problems, the description should be supplemented with pictures, sketches, diagrams and charts. This is to help the user understand how the equipment functions and why it does it in the manner it does.

Installation and Checkout: Specify how the equipment should be unpacked, setup or installed, connected for operation and tested to verify it will operate properly. Although installation and checkout may be very simple, don't overlook the need to warn the user of cautions to be taken during uncrating, lifting, handling and attaching the equipment.

If the nature of the equipment is such that special servicing or adjustment is needed for proper operation, such instructions must be included. These instructions generally must be very specific and be presented in the exact sequence the operations must be performed.

Operation: Explain how to operate the equipment over its range. Specify what must be done, how, and in what sequence.

Tools and Equipment: Identify the tools and equipment that are needed for calibration, maintenance or repair. The items should be defined by manufacturer's name and model number if only certain items can be used. Where generic items are applicable, specify the size or rating so the appropriate items can be found e.g., torque wrench, 150-250 in. - lbs. range. It is helpful to include even common hand tools, since this allows the user to gather the necessary tools in an orderly fashion.

Service and Repair: Describe the steps necessary for the normal servicing of the unit to keep it operating. Such actions as periodic lubrication, cleaning, replacement of batteries, refueling, etc. should be covered.

Repair instructions must be tailored to the complexity of the equip- and the expected capabilities of the person performing the service. There is no point in giving very detailed repair instructions if the user doesn't normally have the skills or equipment necessary to perform these operations. A decision needs to be made early in the engineering process on the levels of maintenance and repair that will be advocated, such as user, service center and factory. Each level requires its own set of instructions, and they should be kept separately.

13.2.3 Guidelines for Writing

Except in large projects where a technical writing staff prepares the instruction manuals, most of the time the task falls on the design engineer. Since technical writing isn't necessarily a strong suit for many design engineers, this section provices some suggestions to make the job a little easier.

The writer, first of all, must understand the theory and functioning of the equipment which is to be described in the instruction booklet. Think carefully about the important features and characteristics that the user should know to use and maintain the item properly. Make a list of these points and use it as a memory jogger when preparing the instructions.

Focus on the user. Recognize that the persons who will use the item know far less about its design and operation than does the designer. Also, initially they tend to skim over the instructions lightly. Consequently, choose words and construct the sentences to convey the meaning clearly. The guiding principle is to keep it simple and direct. Most newspapers about written at about the eighth grade level of grammer and vocabulary. Even still, a large share of their readers still have trouble understanding what is written. Therefore, don't assume the persons reading the instruction booklet will be significantly better than that.

When writing the introductory material (e.g., purpose, general description and theory of operation), use a style which is similar to how you would explain the equipment to an intelligent, but non-technical, neighbor. Use concrete words that have only one meaning in this context. Give brief explanations as you go along to aid the reader's understanding.

When preparing the instruction sections, such as installation, operation and maintenance, use action statements in the form of an order or command, for example:

1. Set throttle level at the START position.
2. Press STARTER button and hold until engine begins running and then release button.
3. Increase engine speed by moving throttle to RUN position.

This style of writing may seem to read a little abrupt, as compared to reading a novel, but it is an effective way of writing instructions. Additional tips for writing are presented in Table 13.1.

Instruction booklets also need good illustrations to aid clarity and understanding. Several different types can be used, and each type is particularly suited for certain applications as noted in Table 13.2. Illustrations should be used generously in conjunction with the text.

TABLE 13.1 Guidelines for Writing Instruction Booklets

Strive for clarity: Use simple words and sentence structures.

Choose words carefully: Select words which have a concrete meaning, not abstract.

Aim the wording at the user: Don't assume the user knows how to operate or maintain the equipment. Help them understand what to do and how to do it.

Use the second person imperative (e.g., Turn selector switch, Press reset button, Insert key, etc.) Avoid personal pronouns (I, we, he, she).

Organize the text logically: Arrange paragraphs in the sequence that the actions normally occur.

Provide roadsigns for the readers: Use headings and subheadings generously so reader can find the desired section or follow the text easily.

Use descriptive headings: Let headings clearly announce what is in the text that follows. Avoid generalities.

Be consistent: Describe identical/similar actions in the same way each time.

Tie illustrations to text: Make reference to figures and tables in the related portion of the text to help reader understand the point of the illustration and the significance of that section of the text.

Stop when you are finished: Say what must be said but stay on target. Avoid extraneous discussions.

Tables are good for arranging and presenting a lot of data in a small space. Tables are especially well suited for summarizing features and characteristics of the equipment, such as dimensions, weights, ratings, operating range, power requirements, type of construction, materials, etc.

Another application of tables is for presenting troubleshooting instructions as shown in Figure 13.1. Other types of instructions can also be arranged in tabular form.

When presenting operating or maintenance instructions, emphasize the correct and safe way to use the equipment. Always use the nomenclature for the equipment exactly as it appears marked on the equipment to avoid confusion. If the lever is marked SPEED CONTROL, don't refer to it as the throttle. Call it the SPEED CONTROL lever. Use caution and warning notes to advise the user of all hazardous conditions that are expected to be encountered. Caution notes must be listed prior to an action which could result in equipment damage. Warning notes also must be presented ahead of any step which could present a hazard to personnel. Such notes should be inserted boldly in the text as shown below.

CAUTION Use unleaded gasoline only. Other types of fuel may damage the engine.

WARNING High voltage can cause electrical shock. Do not remove cover unless power cord is disconnected from electrical source.

It is a good practice to include an explanation in the precautionary notes to assure the reader understands the reason for the caution.

Troubleshooting instructions are probably the most difficult to write. Part of the problem is equipment complexity, and part of it is the many and varied combinations of conditions and events that the equipment can encounter in real life. Thus, the writer must focus on those problems which are the most likely to happen. Beyond that, the section on Theory of Operation must be sufficiently detailed and complete to give a skilled user or knowledgeable technician the ability to apply logic and reasoning to deduce the source of the problem. A description of mechanical relationships or diagrams of information or signal

TABLE 13.2 Types of Illustrations for Instruction Booklets

Photographs: Easy to obtain and relatively inexpensive. Especially good for use in introductory sections of the instruction booklet (description of equipment). Select views which aid understanding. Avoid cluttered or confusing backgrounds in pictures.

Line drawings: Useful for showing dimensions, overall arrangement and placement of components. Often can be adapted from existing engineering drawings. Can be used in introductory sections.

Picture sketches: Especially good for illustrating actions to be taken. placement of tools or equipment, positioning of people or fixtures during installation, disassembly, repair, etc. Generally easier to show what you want in a sketch than in a photograph.

Exploded or cutaway views: Special types of sketches or drawings that are particularly effective for showing interior or hidden details while providing relationships to the overall arrangement. Exploded views can also show assembly sequence. However, these types of drawings are more expensive to make.

Block diagrams/flow charts: Very helpful for showing information or signal flow. Should be used in conjunction with sections on Theory of Operation and Troubleshooting.

TROUBLESHOOTING CHART

Problem	Possible cause	Corrective action
Engine won't start	Out of fuel	Fill tank if empty
	Throttle not in START position	Move throttle to START position
	Defective spark plug	Replace spark plug
Hard to start	Carburetor out of proper adjustment	Adjust carburetor
	Dirt in fuel	Drain fuel tank and refill with clean fuel
Idles rough or stalls frequently	Fouled spark plug or gap setting too large	Clean spark plug. Check gap setting and adjust as needed
	Dirt in fuel	Drain fuel tank and refill with clean fuel
Engine overheats	Low oil level	Fill crankcase to specified level with 10W30 oil
	Cooling air flow blocked	Clean grass or dirt from around cooling fins

FIGURE 13.1 An example of using a table for presenting troubleshooting instructions.

flow should be included. Emphasize what is happening, where it happens, and when it occurs. Describe how and when each function is accomplished so the user or technician can isolate the problem and find the source of the trouble.

Writing instruction booklets is not easy, but it can be done effectively if the writer concentrates on the basics of the equipment and uses a simple, direct writing style.

13.3 REPLACEMENT PARTS LISTS

Although the replacement parts list is similar in content to an engineering/manufacturing parts list, it has a different function. It is a building block in the logistics support system. The replacement parts list identifies those items that are to be used in maintaining or repairing the equipment. This list must be consistent with the level of detail of replacement parts that is stocked by dealers or parts centers. Many engineered items can best be maintained and repaired at the replaceable module or subassembly level. On the other hand, a manual for an overhaul or service depot will have to contain parts lists which identify all pieces to the nut, bolt and cotter pin level.

Replacement parts lists are generally incorporated directly into the instruction booklet for convenience. However, some equipment may be sufficiently complex that the parts list is contained in a separate document. If it is separate from the instruction booklet, the parts list document must have its own identifying document number, and it must be referenced in the instruction booklet.

Presentation of the parts information must be made in a manner which is easily understood by the user or maintenance technician. Some type of picture or sketch is needed to locate and identify the various items. More illustrations are needed as the number and complexity of parts increases.

The most useful type of illustration is the exploded assembly diagram. Although it is the most difficult and expensive to prepare, it presents the information in a clear and convenient manner. The exploded assembly drawing is frequently drawn as an isometric or as a one-quarter front view. The parts and pieces are arranged in various series strings to show where each piece is located in the exact sequence of assembly as illustrated in Figure 13.2. Photographs or simple line drawings can also be used for part identification.

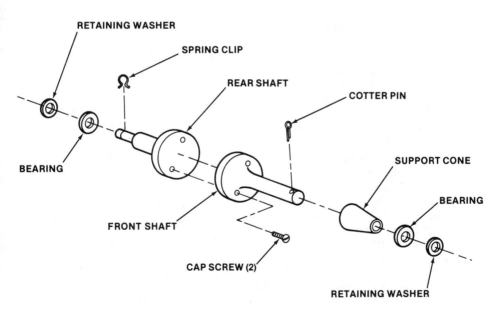

FIGURE 13.2 A typical exploded view.

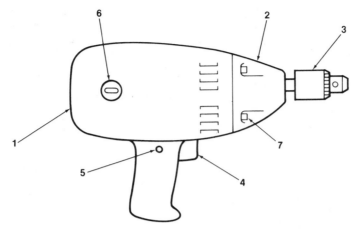

ITEM	NAME	PART NO.	QTY.	ITEM	NAME	PART NO.	QTY.
1	MOTOR HOUSING	16-2114	1	5	LOCK BUTTON	10-6162	1
2	GEAR HOUSING	16-2119	1	6	BRUSH CAP ASSY.	54-0130	2
3	CHUCK ASSY.	HO-3078	1	7	CAP SCREW	08-1942	4
4	TRIGGER SWITCH	63-1590	1	8	OWNER'S MANUAL	OM-16	

FIGURE 13.3 A typical assembly drawing with parts identifiers.

Construction of the listing of replacement parts is also important. The list should be arranged logically so that related parts of an assembly are grouped together and related assemblies are in logical proximity. The listing should begin at the top or highest level of assembly, and proceed downward in a orderly manner to subassemblies and individual parts. Numbers, letters or symbols are assigned to each replaceable piece or assembly, and these identifiers are marked on the photograph or drawing to identify the corresponding item of hardware as shown in Figure 13.3.

For convenience and clarity, it is often necessary to subdivide a large assembly into major subassemblies and present these as separate pictures and lists. For example, the parts list for a portable, diesel-driven, air compressor might be subdivided into six subsections: the engine assembly, compressor assembly, air tank and plumbing system, control panel assembly, and trailer and housing assembly. Each subassembly would then be defined by its own drawing (or sketch) and associated list of replacement parts.

Remember that the accuracy of the illustrations and part identification is very important to assure the user orders and obtains the correct replacement parts. Otherwise, the inherent quality of the design will be degraded.

13.4 DESIGN CONTROL MEASURES

Before instruction booklets and replacement parts lists are issued for customer use, it is important that the documents are adequately reviewed. Instructions should be verified to see if the information is clear and correct. The supporting documentation should be checked carefully against the latest engineering drawings to see that the information is compatible with the design.

Another effective checking method is to have a service technician or factory employee perform the actual operations on a real piece of equipment, using only the instruction booklet for direction. The responsible engineer should be present to witness this activity, hear the comments or suggestions offered by the person performing the task, and note the areas in the instructions requiring additions or corrections. Pay particular attention to actions which may be hazardous and need further precautions specified. This approach is a simple, but effective, way of examining the adequacy and correctness of the instructions.

Parts lists and the associated drawings or sketches need to be checked for accuracy. It is important that the correct part numbers are used and that the items are clearly and correctly identified for the user. Watch for transpositions of letters or numbers, missing or extra digits, leader lines connected to the wrong item, items not shown or not identified, etc. Otherwise, the user will order the wrong parts or be delayed and aggravated while trying to find out how to order the needed items.

After the verification steps have been completed, the final documents should go through a formal review and signoff prior to issue. It is a good practice to have engineering, field service (or manufacturing) and marketing personnel to participate in the final signoff.

Each instruction booklet and separate parts list document should have its own unique identifying number and revision or date of issue. This information is needed for reference and document control purposes. If the replacement parts list is separate from the instruction booklet, the two documents should cross-reference each other for clarity and convenience.

Finally, significant changes to the product design must be incorporated into the instruction booklets and parts lists in a timely manner. Unfortunately, this aspect is often overlooked in the process of evaluating and approving engineering changes. The Configuration Control Board or other change control system (Ref. Chapter 5) should routinely consider the impact of the product change on the supporting documentation. The evaluators of the change must take the necessary actions to maintain compatibility between the hardware and the software.

13.5 SUMMARY

The instruction booklets and replacement parts lists are frequently the manufacturer's only means of advising the user of how the equipment should be operated and maintained. The adequacy of this information often has a direct effect on the reliability of the equipment and on the manufacturer's costs and reputation.

Incomplete or unclear instructions can undermine the effectiveness of the product design. Consequently, the engineering organization must recognize a responsibility for the technical adequacy of the supporting documentation. Whether it is in the preparation or just in reviewing and approving the work of a technical writing group, this responsibility needs to be exercised with the same care and thought that went into the design. The final test of the product is performed by the customer. Be sure your supporting documentation contributes to maintaining the original level of quality of your design.

14

Reliability Improvement

14.1	Introduction	233
14.2	Reliability Reporting System	235
14.3	Measuring and Analyzing Product Reliability	238
	14.3.1 Analysis Tools	239
	14.3.2 Failure Analysis	241
	14.3.3 Corrective Action	244
14.4	Reliability Design Tools	244
	14.4.1 Some Basics	245
	14.4.2 Reliability Reviews	246
	14.4.3 Failure Mode, Effects and Criticality Analysis	246
	14.4.4 Reliability Modeling	252
14.5	Reliability Testing	254
	14.5.1 Test Planning and Analysis	254
	14.5.2 Component Testing	255
	14.5.3 Systems Testing	255
	14.5.4 Reliability Demonstration	256
14.6	Reliability Assessment	256
14.7	Summary	257

14.1 INTRODUCTION

Since the early 1950's reliability has become a significant performance measure. With the increase in technical complexity and the escalating costs of labor and material, unreliable products have a high cost of ownership. It is very important to have a product that works as designed when needed. Thus, nearly all designers must be concerned with the art and science of producing reliable products.

Reliability is commonly defined as "the probability that a product will perform its intended functions under stated conditions for the specified time." It is a probabilistic concept, and the reliability of a product can range from 1.0 (perfectly reliable) to 0.0 (totally unreliable). From the definition, it is obvious that reliability is a function of its planned use, its environment, and time. Each of the parameters has a direct impact on the probability of success.

Reliability is basically dependent upon design. Material quality, manufacturing quality, testing, packaging, handling, installing, operating and maintenance can have a degrading effect on the reliability of the product, but for all intents and purposes, none of these factors can improve the inherent reliability of the design.

For most mechanical and electrical products, their reliability tends to follow a definite pattern. Initially, a new product often exhibits a declining failure rate as initial problems are discovered and corrected. Then for a substantial portion of the product's life, the failure rate is relatively constant. Finally, as the product begins to wear out, its failure rate increases until the product is modified or replaced. This cycle is described as "the bathtub curve" and is illustrated in Figure 14.1.

One of the important concepts for design engineers to recognize and use is the concept of reliability growth. As a product matures, there is usually a gradual increase in the average time between failures. This continues as failure modes are identified and removed or avoided by redesign. Figure 14.2 portrays this phenomenon graphically.

The designer needs to monitor this growth to see if the product is, in fact, reaching an acceptable reliability level for the intended

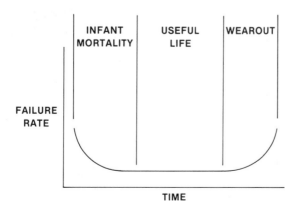

FIGURE 14.1 The Classical bathtub curve.

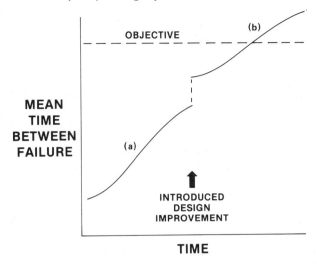

FIGURE 14.2 Reliability growth curves.

application. Curve (a) in Figure 14.2 shows a pattern which is level-ing off ("topping out") below the desired reliability level. Curve (b) shows a pattern of growth which will meet and exceed the stated relia-bility goal as the result of a design change.

There are numerous tools that have been developed by reliability specialists over the years. Some of these tools are reasonably easy to use, and some are quite specialized and require care and specific train-ing to apply. MIL-STD-785 explains many of these methods.

This chapter describes several of the reliability tools that the de-sign engineer can easily use in the day-to-day design activities. The information first presented describes those tasks which address relia-bility improvement for existing or mature products. These include techniques for gathering data, analyzing it to determine the current levels of reliability, and then developing plans and actions for correct-ing the problem areas.

The second set of tools is associated with the design and develop-ment of new products. Included among these are techniques for relia-bility modeling, reliability apportionment, and reliability reviews. In addition, those aspects of testing which are specifically directed at re-liability are presented. Finally, a method for assessing the level of re-liability of a product is discussed.

14.2 RELIABILITY REPORTING SYSTEM

The starting point for all reliability improvement is to know at what level you are starting from. To accomplish this, data on product

performance must be gathered and analyzed. Then, conclusions can
be drawn about the existing level of reliability and whether or not it
is acceptable.

Gathering the data can be aided substantially by developing and
implementing a reliability reporting system.

An effective reporting system provides data in a form which can
readily be used for reliability analysis. However, it requires thought
about what is needed, where to get it, and how to record it for
reporting.

Data which is useful for reliability improvement comes from several
sources. These include field feedback, factory tests, engineering
tests, and any special inspections or audits performed on products before shipment to the customer. The sources listed in Table 2.1 in
Chapter 2 can also be used as reliability input.

Three types of data are especially important for evaluating product
reliability. These are field (operational) failures, service life with or
without failure, and results from engineering tests. Operational failures are significant because they represent experience from the real
world. However, the exact operational and environmental conditions
that were in effect before and at the time of failure may not be known
or reported accurately. Nevertheless, as much data as reasonably
possible should be obtained for each operational failure.

Service life is needed to assess the time aspect of reliability. How
many units are in service? For what period of time? Under what conditions? How many failed and when? Answers to these questions provide considerable insight into product performance.

Finally, results from engineering tests are particularly useful,
since many of the parameters and test conditions can be controlled, or
at least are known. As a result, this data is normally considered to be
dependable for analysis purposes.

To be useful, the data needs to be recorded and submitted to the
persons concerned about reliability; in particular, the reliability engineering group and/or the product design engineers. It is a good
practice to use standardized forms for reporting problems or failures
from the field and the factory. A typical form is shown in Figure 14.3.
This report defines what occurred and provides a written record of
the conditions present at the time of failure or when the problem was
noted.

As each report is prepared, it should be submitted according to a
standard distribution to get the information to the key technical people
who need to know of the problems. Figure 14.4 illustrates the flow of
information for reliability problem reports. The system should be tailored to the size and needs of the organization. Even if the engineering department is very small, it is still desirable to record each reported problem or failure. To accomplish this, the engineer should
write down the failure data and information about problems reported
verbally. Although this is a very informal approach, it is better than
having no data, or trying to carry it all in the engineer's memory.

REPORT NO.	
	R371

PROJECT / MODEL	PART NAME AFFECTED	DATE PROBLEM FOUND
262R79	Filler Valve	5/11/82

PART NO. AFFECTED	NAME OF MAJOR COMPONENT AFFECTED
4025236	Ice Maker

DESCRIPTION OF PROBLEM (WHAT, WHERE, WHEN, HOW MANY, ETC.)

Filler valve does not shut off completely after filling ice making tray and spills water into the ice cube basket.

IMPACT / EFFECT / CONSEQUENCES OF PROBLEM

The spilled water eventually causes the loose ice cubes to freeze together. The frozen mass continues to grow until it completely interferes with ice making.

APPARENT CAUSE OF PROBLEM

Defective seat in filler valve or clogging of valve due to contamination.

REMARKS

May be caused by dissolved salts in incoming water.

REPORTED BY: a g Hyder DATE: 5/11/82	REFERRED PROBLEM TO: Refrigerator Design Group

FIGURE 14.3 Reliability problem report.

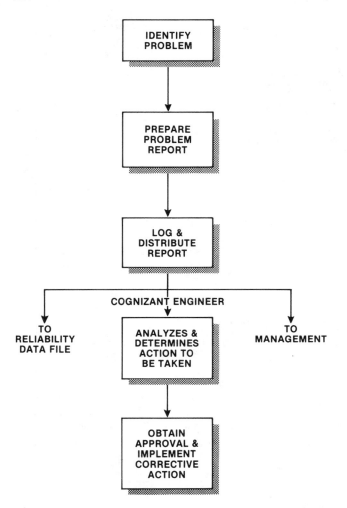

FIGURE 14.4 Flow of information for reliability problem reports.

The reported conditions are then analyzed to determine what action is needed to correct this problem.

14.3 MEASURING AND ANALYZING PRODUCT RELIABILITY

Periodically, it is necessary to gather the data together and examine it for patterns or trends. Several techniques are available for assisting in the analysis process, and these are described in the subsequent sections.

14.3.1 Analysis Tools

There are many tools and techniques available for analyzing reliability problems. Some very sophisticated techniques are presented in the reliability literature such as Weibull plots, hazard analysis and hypothesis testing, but many of these require special knowledge or training to use correctly. But fortunately, there are also several easy-to-use techniques that can be applied to many types of problems. Some of these are described below.

Pareto Analysis. Many years ago it was noted that a small percentage of the items or activities cause a large share of the problems experienced. This is illustrated in the "80-20 Rule," i.e., 80 percent of the problems are caused by 20 percent of the items. Although it is not a precise relationship, it is representative of most situations found in industry.

One of the techniques which is based on this observed phenomenon is the Pareto Diagram. The Pareto Diagram is a pictorial representation of the problems arranged in declining frequency of occurrence or have the largest impact. This is displayed in Figure 14,5. It typically

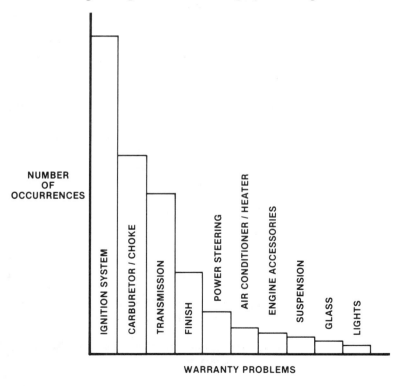

FIGURE 14.5 Pareto diagram.

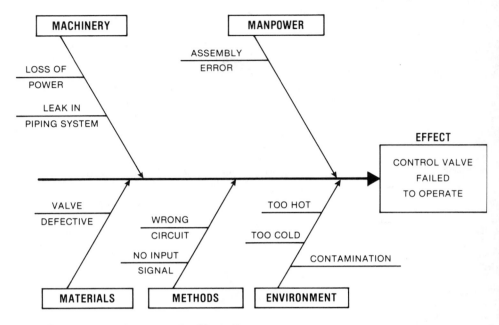

FIGURE 14.6 Cause and effect diagram.

shows which two or three problems are causing most of the trouble.
If corrected, there would be a big reduction in the total number of the
problems.

　　To construct a Pareto Diagram, compile all of the problems occurring during the period of interest. Sort the problems by type (or
cause or dollar effect if known). Group by problem category and arrange in descending order, e.g., largest quantity to least quantity.
Plot the data in bar chart or histogram form. This diagram is particularly useful when presenting the data to management or to persons in
other departments.

Cause and Effect Diagram

The Cause and Effect Diagram is a pictorial means of organizing a problem into logical parts as a problem-solving tool. The diagram, as illustrated in Figure 14.6, shows how the various causes and effects are
related. Although the diagram may not solve the problem, it often
serves as a stimulus which leads to a solution.

　　To construct a Cause and Effect Diagram, first identify the problem and record it in the right-hand box on the diagram. State it succinctly in a few words. Then identify possible causes of this condition.
Frequently, the causes come under the broad categories of Manpower,

Machinery, Materials and Methods. These can be used for starting
the thinking process, and other causes can be added as they are
identified.

Next, specific causes should be listed under each of the appropriate broad categories. For example, under "Manpower," the contributing causes might include: human error, incorrect instructions, lack
of training, etc. Under "Materials," the cause could include: wrong
material, defective material, material too hard, etc.

From this, the analyst then can pursue those causes which appear
to be major contributing factors. This requires getting additional information to determine whether or not this is an item which causes the
identified problem (effect).

The Cause and Effect Diagram can be applied to many reliability
problem-solving situations.

Mean Time Between Failure (MTBF): This is a numerical value
which represents the average operating time between failures. It is
calculated by dividing the total cumulative amount of operating time by
the total number of failures (typically expressed in hours, but it
can also be in other units of time or operating cycles). It is a commonly used measure of reliability—the larger the number, the more reliable the product.

The reciprocal of MTBF is failure rate, which is another important
parameter for measuring the reliability of a product. Thus, knowing
one value allows the quick calculation of the other. Failure rate is
typically expressed in terms of failures per unit of time.

Chi Square Method: Many times there is only scant failure and
life information available and simple analytical calculations (e.g.,
MTBF) do not give statistically reliable results.

In these situations, the "Chi Square" method is useful. Only two
pieces of data are required: number of failures and the total amount
of operating time. Tables 14.1 and 14.2 explain how to perform the
calculation. This data provides an insight into the estimated reliability
early in the product's life.

14.3.2 Failure Analysis

As each failure is reported, it should be investigated to determine the
mode of failure and its cause. The results of the investigation should
be documented in some form of failure analysis report. It may be a
one-page memo, referencing the original failure report and describing
the conclusions of the analysis in a couple of paragraphs, or it may be
a comprehensive formal engineering report.

In the event that it is not possible to determine the cause of failure, a memo should still be issued, describing the conditions and noting the cause of failure was not identified. All such memos should be
filed together and periodically reviewed in light of new information
that comes available.

TABLE 14.1 Estimate of Reliability Using Chi Square Method

$$MTBF = \frac{2T}{F_N}$$

MTBF Mean time between failures, (See T for units of measure)

   T Total operating time of the population of interest (in hours, days, years, cycles, etc.)

 F_N Chi square factor (See table 14.2)

   N Number of failures in population of interest

Note: Calculation is normally performed for the 90% & 10% confidence levels.

   Example: A batch of 30 prototypes of a new vacuum pump were built and tested. There were a total of four failures by the time the group of pumps had operated for a cumulative total of 7800 hours. The mean time between failure is estimated as:

$$MTBF_{90\%} = \frac{2\,(7800)}{F_{4(90\%)}} = \frac{15,600}{4.17}$$

$$MTBF_{90\%} = 3741 \text{ hours}$$

$$MTBF_{10\%} = \frac{2\,(7800)}{F_{4(10\%)}} = \frac{15,600}{14.7}$$

$$MTBF_{10\%} = 1061 \text{ hours}$$

With 90% confidence, the mean time between failures will be greater than 1061 hours and not more than 3741 hours.

   Failure analysis is not an exact science, but advanced technology now makes it possible to pursue avenues of investigation heretofore not available, such as inspection by electron microscope, mass spectrometry, or x-ray diffraction. But this work must be done by specialized materials laboratories. Even though not every company can afford a laboratory equipped with such exotic devices for failure investigation, most companies can still perform some physical tests and examinations. Another possibility is to subcontract such investigations to a nearby independent laboratory, a local university or perhaps a large neighboring industrial firm that has laboratory facilities. In any event, some effort should be made to determine the mode of failure and its cause.

Photographs should be taken of the failed item and of any subsequent tests or experiments performed on it or on similar items. This provides a convenient and inexpensive record of the conditions observed and is often revealing when viewed by other knowledgeable persons.

Wherever possible, it is a recommended practice to perform tests to see if the failure condition can be duplicated. If the failure can be caused to occur in a controlled test, it provides a strong assurance that the mechanism of failure is truly understood. Then appropriate action can be taken to avoid this problem in the future.

TABLE 14.2 Chi Square Factors (F_N)

Number of failures	Confidence levels	
	90%	10%
0	0.0158	2.71
1	0.584	6.25
2	1.61	9.24
3	2.83	12.0
4	4.17	14.7
5	5.58	17.3
6	7.04	19.8
7	8.55	22.3
8	10.1	24.8
9	11.7	27.2
10	13.2	29.6
11	14.8	32.0
12	16.5	34.4
13	18.1	36.7
14	19.8	39.1

NOTE: Table based on chi square values
for various degrees of freedom (DF)
(DF = 2 × no. of failures + 1)

14.3.3 Corrective Action

It is the responsibility of the design group to decide what action is
needed to prevent recurrence of the problem. In small firms, this entire process may involve only one or two persons. The tasks can be
handled quickly and informally. However, in larger organizations,
several different groups may be involved in the identification, investigation and resolution of the problem. In those organizations which
have a separate reliability engineering group, it is common for the reliability engineer to be in charge of the data gathering and failure investigation. When the failure analysis is completed, the reliability
engineer then issues a Request for Corrective Action, a Reliability Action Request, or similar-titled memo to the cognizant design group to
request a design change to resolve the problem. With such systems,
it is the usual practice to maintain a log of corrective action requests.
Periodically, status reports are then published to show which requests
are still open and those which have been implemented.

Another good engineering practice is to monitor the implementation
of design changes which were specifically intended to resolve a failure
condition or recurring field problem. The purpose of this follow-up of
the revised design is to verify that the corrective action fixed the
problem.

Monitoring should be continued for a sufficiently long period to allow the modification to experience actual service conditions. Six
months to a year are typical monitoring times, but it may vary with
the application. Results from the field feedback should be summarized
and reported periodically (e.g., quarterly) to the engineering and reliability personnel to show what is happening.

If the results show a definite improvement pattern during the pre-
determined monitoring period, this type of special tracking can be discontinued. If there isn't an improvement, different corrective action
will be needed.

Don't assume that the corrective action automatically resolves the
problem every time. Check and find out.

14.4 RELIABILITY DESIGN TOOLS

Up to this point, the emphasis has been on working to improve the re-
reliability of existing, or even mature, products. Although the tools
described so far can be applied to new products also, they tend to be
more related to the operational phase. However, there are also a number of reliability techniques that can be readily applied during the design phase. These are defect prevention tools which focus on getting
it right from the beginning. As such, they also play an important role
in design assurance. The reliability techniques presented in this section are easy to use and can be applied to nearly every type of product.

Although there are other more sophisticated methods available, the emphasis in this book is directed at the fundamental tools. The thought being as the readers try these and gain both success and confidence they will be properly motivated to seek out the more advanced methods and learn how to apply them to their problems.

14.4.1 Some Basics

One of the guiding principles of designing for reliability is: Keep it simple. Simplicity is the keynote.

In many situations, engineering departments have added parts and functions to their products to achieve greater technical performance, only to find a noticeable drop in the reliability of the product. The added complexity provides more opportunities for something to go wrong. And, unfortunately, it frequently does.

Another of the techniques for improving reliability is to build in redundancy. Redundancy can be defined as the existence of more than one way of performing a given function. This typically is implemented by having various parts (not necessarily the same) in parallel, rather than in series. Thus, only certain of the parts are required to accomplish the intended function. However, even redundancy must be used with caution, because it is contrary to the principle of keeping it simple.

Redundancy is frequently used successfully in electrical/electronic equipment and structural elements, and to a lesser degree, in mechanical devices. But there are other aspects of redundancy that must be considered. Additional parts or systems usually add cost, weight, physical size and operational sequences. Further, parallel parts or systems may not really be redundant, because similar parts typically have common connection points and common failure modes. As such, the same fault condition may adversely affect all redundant elements simultaneously. But with intelligent design, redundancy can frequently improve the operating reliability of a component or system. It is a technique to consider.

Another valuable technique for improving reliability is to use proven parts in the design. This is generally accomplished by compiling a special list of those parts which have been used successfully in service and have a low or zero failure rate. Designers are then expected to use these parts in their designs in preference to new or unproven parts. It is another approach to standardization which is also desirable from a reliability standpoint.

A parallel to the use of proven parts is the use of proven design methods. This was described in some detail in Chapter 6 but is repeated here because of its significance as a tool when designing for reliability.

Another reliability design technique which is frequently used in mechanical, structural and electrical design is that of derating.

Although a material or component may be rated for successful operation at a given load and/or temperature, it is intentionally applied to
operate at a lower load or temperature condition. By operating at a
less severe level, the item is then expected to perform more reliably.
However, it should be noted that there are tradeoffs in derating that
must be considered, such as increased initial cost, greater weight or
volume, and possibly greater energy requirements. Nevertheless, derating is an appropriate design technique for many applications.

14.4.2 Reliability Reviews

In Chapter 9, both formal and informal design reviews were identified
as valuable design assurance tools. Representatives from various technical and non-technical disciplines serve as the review team members
and provide constructive inquiry to aid in achieving a well-designed
product. Similar reviews are also used to enhance product reliability.
In reliability reviews, the design and design approach are explored to
see if those aspects important to reliability have been properly and
adequately addressed. Table 14.3 provides a series of questions to be
asked and answered during the Reliability Reviews.

These questions can be used in both formal and informal reviews.
Also, it may be necessary to ask them at various points in the design
process, such as preliminary design, intermediate design and final design. These same questions can be applied in, perhaps, modified form
by the reliability representative on the Configuration Control Board or
when reviewing proposed design modifications to existing products.
Reasonable answers to all such questions should be provided by the
designer.

Another output from each reliability review should be a risk assessment. Areas of concern or higher-than-usual risk should be identified by the reliability representative to local management. In fact,
it is a recommended practice to publish a list of risk areas and review
the status of each item at each subsequent reliability review. There
should be visible progress of resolving the concern and reducing the
level of risk as a result of further analysis, testing, design changes,
etc. Otherwise, the reliability representative should blow the whistle
even louder!

14.4.3 Failure Mode, Effects and Criticality Analysis

There is another analysis technique that is valuable for identifying and
assessing weak links in the product design. It is known as Failure
Mode, Effects and Criticality Analysis (FMECA). This technique was
developed and honed in the aerospace industry and was used by NASA
in their efforts to make reliable space vehicles.

Even though it began as a design tool for advanced aircraft, it can
be used effectively on virtually any product. And it is an analysis

TABLE 14.3 Reliability Review Checklist

1. Is this a new design or new application of an existing design?

2. How is this design similar to and different from the company's existing products (or design experience)?

3. What problems/failures have been encountered with similar designs?

4. What has been done in this design to avoid these problems?

5. What are the loads/stresses for:
    (a) startup (b) standby (c) normal operation (d) emergency operation (e) shutdown (f) shipping (g) storage (h) other important conditions (specify)

6. What are the environmental conditions for:
    (a) startup (b) standby (c) normal operation (d) emergency operation (e) shutdown (f) shipping (g) storage (h) other important conditions (specify)

7. What are the most severe (limiting) combinations of conditions pected? (load/stress/environment)

8. What is the design margin for each of the limiting conditions (allowable stress minus expected stress)?

9. What parts/assemblies are new or have not been used operationally in similar products?

10. What features are included in the design to enhance operational reliability?

11. What are the most likely failure modes?

12. What has been done to reduce the risk of these occurring?

13. What is the expected reliability (probability of success)?

14. How does the this compare to the required level of reliability?

15. What are the areas of greatest risk or concern in this design?

---

that can readily be done by the designer. However, it is a preferred practice for the designer and the reliability engineer to work together on it.

A detailed explanation of the FMECA method is contained in MIL-STD-1629. The methodology is summarized in the subsequent paragraphs.

Developing a FMECA is normally a bottom-up process, although it can be applied at any level of assembly. At the selected starting level, the engineer examines each item and determines how the component can fail. Each failure mode typically has a different effect and must be considered separately.

When considering what failure modes can occur in a component, there are at least four that should be investigated. These are:

1. Failure to function at the prescribed time
2. Failure to cease functioning at the prescribed time
3. Premature operation
4. Operate in an incorrect manner

For any one failure mode, there can be several different causes for failure. For example, a hydraulic actuator could fail to perform its function due to a seal leak, a rupture of the cylinder, binding of the shaft, breaking a link pin, loss of supply pressure, to name a few.

After identifying the various failure modes, the analyst must determine the effect of each failure mode on the operation of the component, on the higher-level assemblies, and finally on the functioning of the end item. Next, the analyst should estimate the likelihood of such an event happening during operation. Either a quantitative or qualitative estimate can be made. Obviously, a failure mode which has a serious adverse effect on the system and has a high probability of occurrence is reason for concern and is a candidate for corrective action.

To further evaluate the significance of the failure condition, the analyst should also determine the method and relative ease of detecting a pending failure before it occurs. This may make it possible to mitigate the failure through instrumentation, maintenance action, redundancy or similar methods.

One of the rules for performing a FMECA is to treat each failure as a singular event, i.e., it is the only failure condition present in the system at that instant. It is appropriate to identify subsequent failures (domino effect) which are caused by the original condition. But do not try to evaluate the effects of multiple failures simultaneously. It generally is too speculative to draw meaningful conclusions.

The analysis process needs to be recorded, so it can be reviewed and acted upon. Although the exact format is not critical, the FMECA form should contain space for listing the component, its failure mode, the effects of the failure at two or three higher levels of assembly, the means and ease of detecting pending failure or compensating for it, the likelihood of failure, and the criticality of the failure. Figure 14.7 shows a representative form that can be used for failure mode analysis.

Finally, the analyst must draw conclusions about the potential failure conditions. One way to approach this is to use a Failure Criticality Index, such as explained in Table 14.4. The factor weightings are somewhat arbitrary and can be adjusted to fit local preference. However, the use of weighting factors does help in prioritizing the potential failure conditions. This is an important step, since the final

SYSTEM: MAIN GENERATOR				DATE 12 / 7 / 83		
COMPONENT / SUBSYSTEM: COOLING SYSTEM				SHEET __3__ OF _9_		
IDENT. NO.	FUNCTION	FAILURE MODE	EFFECT	EASE OF DETECTION	PROBABIL. OF FAIL.	CRITICAL-ITY
4-1	COOLANT PUMP DELIVERS CONSTANT VOLUME OF COOLING WATER	SHAFT SEIZURE DUE TO BEARING FAILURE	LOSS OF FLOW	DIFFICULT UNLESS MONITORS ADDED	MODERATE	SERIOUS
		PUMP MOTOR WON'T RUN DUE TO LOSS OF ELECTRICITY	LOSS OF FLOW	EASY	LOW	SERIOUS
4-2		REDUCED FLOW DUE TO CORROSION OF PUMP IMPELLER	NO EFFECT UNTIL FLOW FALLS BELOW MIN. REQ'D VALUE	MAY NEED CONTIN. INCR. IN FLOW THRU CONTROL VALVE	LOW	LOW
	CONTROL VALVE MODULATES FLOW OF COOLING WATER TO HEAT EXCHANGER BETWEEN MAX & MIN. VALUES	FLOW REDUCED BELOW MIN. DUE TO ORIFICE BLOCKAGE	OVER-HEATING OF MAIN GENERA-TOR	EASY	LOW	SERIOUS

FIGURE 14.7 FMECA worksheet form. (From: J. A. Burgess, Spoting Trouble Before It Happens, *Machine Design*, Sept. 17 1970, Penton/IPC, Cleveland.)

outcome of a FMECA is to decide what actions need to be taken to improve the reliability of the component and the system.

After completing the FMECA forms for all of the components in the assembly or the system, a summary listing should be prepared to show the failure modes which pose the greatest risk to system success. This can be done by listing the failure modes in greatest-risk sequence or by listing the components in declining order of the risk weighting factors. Figure 14.8 illustrates one such summary. While constructing the summary, examine the crucial problems identified to see if they seem to be reasonable from an intuitive standpoint. In other words, look again to be sure the analysis has not led to unreasonable—or absurd—conclusions.

The finalized summary of critical failure modes then serves as a priority listing for corrective action. It is important to follow through on the conclusions and recommendations from the analysis. Otherwise, the effort spent on the FEMCAs was wasted. The list should be distributed to engineering and project management for action. It can be

TABLE 14.4 Failure Criticality Index

$$FCI = F_1 \times F_2 \times F_3 \times F_4 \times F_5$$

Class of failure	F_1
Catastrophic	3.0
Serious injury/damage	1.5
Nuisance	0.8

System effect	F_2
Significant effect on two or more higher assemblies	2.0
Significant effect on next higher assembly	1.0
Little effect on higher assemblies	0.5

Likelihood of occurrence	F_3
Definite	1.5
Likely	1.0
Remote	0.5

Ease of detecting impending failure	F_4
Difficult	1.3
Reasonably easy	1.0
Very easy	0.7

Design familiarity	F_5
Considerably different from existing hardware	1.2
Similar to existing hardware	1.0
Uses standard hardware	0.8

FCI values greater than 1.5 generally warrant preventive or corrective action.

---

used for planning special or supplementary tests, further analysis, product redesign or changes in operating or maintenance procedures and practices. The failure modes of concern should also be added to the list of high-risk items and examined periodically in the reliability review process.

CRITICAL FAILURE MODE LIST

Component	Critical failure mode	Failure criticality index	Remarks
Coolant pump	Loss of output due to bearing seizure.	5.3	May need alarm system to alert of eminent failure.
Control valve	Orifice blockage	2.7	Should add filter or screen upstream. Will require scheduled maintenance to clean screen.
Control valve	Loss of acuation power	1.9	Need to examine alternate power sources
Heat exchanger	Leaks due to corrosion	1.6	Must control pH of coolant water.

FIGURE 14.8 Critical failure mode summary sheet. (From: J. A. Burgess, Spotting Trouble Before It Happens, *Machine Design*, Sept. 17 1970, Penton/IPC, Cleveland.)

Although the FMECA process is described as a tool for design, it is equally applicable to the engineering change or redesign process. In fact, when a redesign is made, the new design should be evaluated in the same manner as the original design. The analyst can then verify that the old problems were corrected and that no new problems of consequence were created.

14.4.4 Reliability Modeling

Another of the tools of reliability engineering that is useful in most applications is the Reliability Block Diagram. It is a simple picture that shows the relationship of components in a system or of the functions of a given component. This typically consists of a line drawing showing the series-parallel relationships of parts, components or functions. Although it may look elementary when completed, it frequently requires considerable thought to determine the true relationships among the components or of their functions. Figure 14.9 shows an example of such a block diagram.

This schematic is useful in examining the consequence of various failure modes and for determining which components represent the critical paths needed to accomplish the product's intended function. The pictorial representation can show the weak links in the design that should be strengthened by adding redundancy, design margin, special test requirements, etc. Reliability block diagrams also serve as a visual aid during the performance of failure mode, effects and criticality analysis.

Another technique is the mathematical modeling process described in the classical reliability engineering texts. In this methodology, failure probability values for each component are combined mathematically, according to the specialized equations, to calculate an overall reliability value for the complete assembly or system. For large or complex systems, this can be a long and tedious process, frequently requiring a computer program to reduce the time and manual effort. However, it should be pointed out that the absolute values of all such numbers should be viewed with a good bit of skepticism. Even though the resulting number may be mathematically correct, its relationship to the real world is heavily dependent upon the accuracy of the listed failure probability numbers and of the logic and accuracy of the model. Nevertheless, it is often useful for relative comparisons of alternate designs and for identifying those components or functions which appear to be the areas of lowest reliability. It is another tool for planning corrective action.

In some high-technology programs, reliability block diagrams and mathematical models are used to apportion or allocate the overall reliability goal among the constituent components. These values are expressed as specific component failure rates or mean-time-between-failures. The values are normally selected from published data or

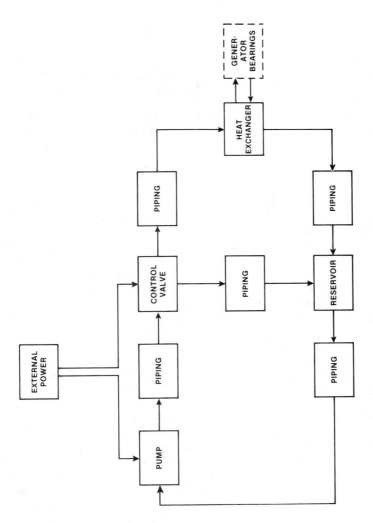

FIGURE 14.9 Reliability block diagram. (From: J. A. Burgess, Spotting Trouble Before It Happens, *Machine Design*, Sept. 17, 1970, Penton/IPC, Cleveland.)

based on local experience with particular components. The reliability
values then are incorporated as design goals in the component and
system specifications. In some situations, the reliability numbers are
defined as requirements, and the manufacturer must later demonstrate
compliance with these values through analysis or test.

14.5 RELIABILITY TESTING

One of the fundamental tools in reliability measurement and improvement is testing. However, as noted in Chapter 8, testing can be expensive, require considerable time and effort, and may be limited due
to facilities or measurement capabilities. Regardless, testing is a major source of data for reliability purposes. But to get the most from
the testing effort, certain actions and techniques need to be applied.
These are described below.

14.5.1 Test Planning and Analysis

The need for reliability testing can come from many sources, such as
engineering judgement, design review comments, failure mode and effects analysis, tradeoff studies, problems occurring on similar projects
or components, results from design analyses, etc. For many products,
the type and extent of the tests may be reasonably obvious and
straight forward. Thus, defining the test requirements may be equally
simple. Nevertheless, it is a preferred design assurance practice to
document the requirements in the form of a test plan or test request.
When the test requirements are more extensive or complex, it is all the
more reason to record them for planning and implementation purposes.

The test request should clearly identify the areas of risk or concern that need to be investigated, in addition to the basic information
specified in Section 8.2 of Chapter 8. Further, the requester should
include a prediction of the expected results of the test. This information will then serve as a guide for test monitoring and is useful as acceptance criteria when analyzing the actual test results.

Although some tests may be less valid than others, due to hardware availability, facilities limitations, etc., nearly all tests can provide some valid data for reliability purposes. However, when there
are limitations or constraints present, these should be declared, and
the estimated impact on the results should be stated prior to the performance of the test. This is to define before the fact which part of
the test results are expected to be invalid and which should be considered valid for reliability purposes. The effect of this declaration
on the acceptance criteria should also be stated.

After each test, the results should be compared with the test predictions and test declarations and conclusions drawn. The results of

the analysis can then be used with reasonable confidence in the compilation of success and failure data for the component or system.

It is especially important to consider all unexpected failures as a warning of a potential problem condition for reliability purposes. Even if its cause seems to be a fluke condition or due to an apparently unrelated cause, it is still a good practice to include it in the failure statistics. Although this may seem overly conservative, the unexplained failures are often a symptom of a real problem which may surface later.

In the event there is a failure, the test group should document the conditions present at the time of failure for inclusion in the test records. This information is very valuable for proper failure analysis. In addition, operating time, cycles or test time should be recorded, since this is needed to evaluate the life aspects in the reliability equation.

14.5.2 Component Testing

Testing of components is an important step in any reliability improvement program. It should be performed early in the cycle, and the tests should be carried to failure. This is to determine design margins and failure modes. Component testing should be used on all unproven components and on proven components which will be applied in a manner different from prior applications. These tests provide a basis for deciding whether or not it is appropriate to commit the component to systems development. Results from component tests should *not* be used for predicting the reliability of the component in systems operation, since it may see other interfaces and environments. Nevertheless, the component test data is valuable for flushing out weaknesses in the components, which eventually could have had an adverse effect on the operating system.

Although component tests can provide basic design data and weed out weak designs, it is not a substitute for systems testing. Therefore, for reliability and design assurance purposes, do not stop with component testing.

14.5.3 Systems Testing

Systems testing, i.e., testing of the complete unit or top assembly, is fundamental to reliability improvement. To the maximum extent possible, the end item should be tested under actual (or closely simulated) loads and environmental conditions. Systems tests are intended to explore the effects of component interactions (including software as applicable) and the real world environment.

For maximum effectiveness, systems tests should be conducted on an iterative basis of test - fail - fix - retest. Although there are differing schools of thought on this, the author firmly believes that the most effective method for improving reliability at the hardware stage

(after doing the best design job possible) is to find the failure mode
for the weakest link, design it out, and then test for the second weakest link and design it out, etc., until there is an adequate level of
margin for the intended application.

14.5.4 Reliability Demonstration

Many of the texts on reliability engineering describe a special series of
reliability tests. These tests are fashioned to verify the life capabilities of the product. As can be imagined, such tests may be very expensive and time-consuming. Consequently, it requires intelligent
planning of the tests to get the maximum information from the fewest
number of tests. It is equally important that the parts to be tested
and the performance of the test be closely controlled to assure the
validity of the results. Regardless of these concerns, special life-
load-environment tests can be very valuable in assessing the reliability
of a design. MIL-STD-781 is a useful reference for such tests.

Some types of products can be subjected to accelerated life tests,
such as testing at elevated temperatures or stresses, to compress test
time. However, this approach is not valid for many products. In
other cases, accelerated testing may distort or hide other conditions
which are equally serious but do not respond in the same manner under
the accelerated test conditions. Therefore, accelerated testing must
be used with caution.

The guiding principle in reliability testing is to plan the tests
carefully. They can provide valuable data, but only if done right.

14.6 RELIABILITY ASSESSMENT

Reliability assessment is a constructive, self-examination process. It
should be done periodically for both new and mature products to determine the trend of product reliability. To do this, the cognizant
engineering or product management group should first define what are
considered to be the key parameters for measuring the reliability of
the product. This typically includes factors such as, total number of
failures, failure rate, warranty cost per unit or as a percent of sales,
customer returns, customer complaints, competitor's relative reliability
performance, field service frequency, etc.

For many products, this information should be gathered and plotted monthly or quarterly. But in most cases, it will be adequate to
examine the data once or twice a year. In any event, the assessment
should consist of compiling the various data, comparing it to pre-established standards or norms, and drawing conclusions about the acceptability of the results. Each assessment should also be compared
to previous evaluations. The purpose is to see if the reliability is
getting better or worse as a function of time.

A recommended practice is to document the results in a letter report to local management. The report should summarize the latest reliability data and show the trend or pattern of the product's reliability for the current and previous periods. This can be done in tables or graphs, although the graphical presentation is the easiest to see and understand. A brief explanation of the data and the conclusions reached on the trends should be included in the report. It is also appropriate to highlight the areas of greatest concern or risk in the report and note what progress is being made to resolve these problems.

Such assessments provide a snapshot of the product's reliability performance. This can serve as a means for monitoring reliability growth and provide an early warning system in the event that symptoms of reliability problems are showing up. For a small investment of time and effort, the periodic reliability assessments can provide important information to engineering and project management.

A similar process can be used as new products are developed and introduced. Although the data base may be smaller, the information and the results from the evaluation are useful in the product development process. The results should be examined at the reliability reviews, and problems noted should be incorporated into the risk monitoring system.

14.7 SUMMARY

The reliability of a product is largely a design-dependent parameter. It is also affected by its use and environment and as a function of time or cycles. Various methods and techniques have been devised for predicting, measuring and evaluating the reliability of a product. However, there is a considerable amount of uncertainty associated with absolute numbers, especially as it applies to reliability predictions. Nevertheless, these tools provide valuable insights for planning and implementing actions to improve the relative level of reliability of a product.

The cardinal rule for reliability improvement is to identify the critical failure modes, find their root causes for occurrence and take action to prevent these from happening in the future. More times than not, this will require design action to eliminate or avoid the troublesome conditions.

Data collection and analysis is the keystone for reliability improvement. Once the data has been compiled and examined, it is often clear what the problem is and what needs to be done to correct it. Occasionally, special statistical techniques may be required to identify the more subtle problems and their causes.

An on-going data gathering process, coupled with periodic evaluations, provides a means for monitoring reliability over a period of time.

The management axiom: "If you can't measure it, you can't manage it," especially applies to reliability.

In the long run, reliability becomes a key characteristic of the product. If the reliability of the product is low, it has serious consequences in the marketplace. If reliability is high, it can easily become one of the most important product features to your customers.

15

Auditing the Engineering Process

15.1 Introduction 259

15.2 Fundamentals of Systems Auditing 260

    15.2.1 Planning and Preparation 261
    15.2.2 Performing the Audit 262
    15.2.3 Reporting the Results 264
    15.2.4 Followup and Closeout 267

15.3 Areas for Investigation 267

    15.3.1 Basis for Design 268
    15.3.2 Drawings and Specifications 268
    15.3.3 Design Analysis 268
    15.3.4 Test Programs 269
    15.3.5 Engineering Changes 269
    15.3.6 Control of Nonconformances 270

15.4 Ethics for Auditor Conduct 270

    15.4.1 Role of the Auditor 270
    15.4.2 Objectivity 271
    15.4.3 Compliance with Local Customs 271
    15.4.4 Impersonal Reporting 272

15.5 Improving the System 272

    15.5.1 Taking Action 272
    15.5.2 Audit Timing 273

15.6 Summary 274

15.1 INTRODUCTION

Auditing was developed many years ago in the accounting field. Its
purpose was to examine the methods used for controlling the finances
and for verifying the accuracy of the financial reports. Audits provide management with a check of the system and give an indication of
how well it is working.

In the past 25 years management recognized that similar auditing
techniques could be applied to many other areas of business. In these
other areas, audits can evaluate the effectiveness of the systems,
methods and actual practices. However, it is only in recent years
that the auditing techniques have been used to examine the engineering process.

An audit is defined as a planned and systematic evaluation of an
activity against specific criteria and requirements. Systems audits
provide answers to management on the following important questions:

1. Do the people understand how they are expected to perform
    their work?
2. Are the necessary procedures available?
3. Are they followed?
4. Are these methods effective in accomplishing the intended
    purpose?

By investigating these questions through the systems audit process, engineering management can frequently improve both the operation of the organization and the workings of the design assurance
program. Internal audits frequently reveal duplication of effort, in-
consistent actions, shortcomings of management controls, unnecessary
procedures, and inadequacies in the information flow system.

This chapter describes the basic elements of an engineering audit
and provides specific guidelines and recommended practices for auditing the engineering process.

15.2 FUNDAMENTALS OF SYSTEMS AUDITING

There are four major steps in conducting an internal audit. These
same steps apply to audits of all types of activities, e.g., engineering,
purchasing, manufacturing, marketing. Only the avenues of inquiry
and the specific questions used are different for the various audits.
The four steps are:

1. Planning and preparation.
2. Performing the audit.
3. Reporting the results.
4. Followup and closeout.

Each of these major steps are described in subsequent sections.

15.2.1 Planning and Preparation

The initial step in the planning process is to establish the scope of the audit. This can be done by answering questions such as: What areas or activities should be evaluated? Should it cover the total effort or just a portion of it? Or perhaps should it examine only a particular segment of the work? Scoping should be done by the manager or executive who wants the audit performed. One approach is to select specific activities to be audited (drawing preparation, stress analysis, design verification testing, engineering change control, etc.). Another approach is to select a specific group to evaluate (design, drafting, test lab, change control board, etc.). This can be done for a specific program or project, or it can be performed across the board (covering several or all active programs). As it will become apparent, it is a significant task to check all of the various kinds of work performed in an engineering department in a single audit. Thus, it is frequently necessary to divide the task into two or more audits, performed over a period of time.

After the general scope has been established, personnel knowledgeable of these functions or programs should be selected and assigned to perform the audit. A group of two or four persons is a good working size for an audit team, although one qualified person may suffice for relatively small or simple audits. Persons selected for the audit team should have some prior training in the art and science of auditing and not try to do it just intuitively. Many good short courses are available in industry, or perhaps the necessary training can be obtained from the company's quality assurance department.

Experience shows that the auditors should not be the ones that performed the work being audited or that were directly responsible for having it performed. However, they should have sufficient knowledge and understanding of the work to judge its adequacy and to determine whether or not the work is being accomplished in a manner consistent with the management policies and requirements. In addition, the auditors must evaluate the effectiveness of the administrative methods used for controlling the task. An experienced quality assurance auditor, plus one or more knowledgeable engineers, make a good internal audit team for an engineering department audit.

The team's first task is the planning of the audit. This includes studying governing requirements, reviewing applicable directives and procedures, and compiling a list of questions to be asked or specific points to be verified. Appendix 3, at the back of the book, presents a list of guidelines for use in an engineering audit. It can serve as a roadmap for planning the audit and for instructing the inexperienced auditors in how to proceed.

One person should be designated as the team leader and be responsible for overall coordination of the audit. Areas for investigation are then assigned among the team members, and a written plan for the

TABLE 15.1 Typical Contents of an Audit Plan

Name of the group to be audited.

Names of audit team members and their organizational affiliation.

Brief statement of the scope of the audit.

Listing of areas or activities to be audited.

List of applicable requirements (standards, regulations, procedures, etc.)

Tentative schedule for the audit.

From: J. A. Burgess, Auditing an Engineering Organization, *Mechanical Engineering*, Sept. 1979, American Society of Mechanical Engineers, New York.

audit is developed. Table 15.1 lists the contents of a typical audit plan. Although some people like to perform unannounced or surprise audits, it is not necessary and frequently produces a feeling of distrust which is counter-productive. The audit team can make a number of random checks to assure that a scheduled investigation produces valid results without resorting to the surprise approach.

Management of the groups to be audited should be contacted to work out the final arrangements for the audit, and the audit plan should be finalized at that time. The team is then ready to begin the audit.

15.2.2 Performing the Audit

At the beginning of the audit, the team members should meet with management representatives of the group or activity that will be audited. This may be the engineering manager or the supervisor of the particular activity to be examined. During this meeting, the purpose of the audit should be explained, and the methods that will be used, such as the review of documents, interviews with members of the group, etc. should be discussed with the cognizant management representatives. It is necessary to establish early in the entrance meeting that the purpose of the audit is to examine the use and effectiveness of the administrative controls and not to find fault with individuals or their specific work. This is especially important to avoid feelings of antagonism or distrust.

To reduce the chances of encountering such feelings, some auditors ask local management if there are any areas they would especially be interested in having the audit team examine as a part of the audit. This often helps to overcome the concerns of the group being audited.

Now the actual auditing can begin, using one or more of the techniques listed in Table 15.2. By applying these methods, the audit team can determine if there are areas where the engineering department is having problems. The auditors may also be able to identify the causes of these problems. However, take care to avoid nitpicking, dwelling on trivia, or concluding there is a need for major change based on only casual observations.

The auditors must also take special care to avoid criticizing a particular individual or finding fault with decisions they have made. Objectivity is the key to this effort. Let the facts speak for themselves.

As the auditors conduct their interviews and examine records, they should be making notes of their investigations. This includes identification of items that were reviewed, such as drawing numbers, calculation sheets for particular items, report titles, etc., and any findings or irregularities, that were noted. If no problems are found in a particular area, it is also a good idea to note this fact and comment to that effect in later reports. Notes taken during the audit serve as the basis for the final audit report. Detailed observations and specifics provide credibility if the auditors' conclusions are later questioned.

At the completion of the audit, an exit meeting with the responsible persons from the activity that was audited should be held. The purpose of this meeting is to convey the results of the audit to the management of the group. Any nonconformances or observations

TABLE 15.2 Techniques for Performing Audits

---

1. Request local management to explain what work is performed in the group and how it is done. Then review actual activities and records to determine if it is really done that way.

2. Compare the performance of work and review existing documents and files against applicable written policies and procedures.

3. Select a contract or project and follow it through the various phases to determine if it is being accomplished as required by the contract requirements and management directives.

4. Examine complaints, returned products or failure data against expected performance norms and look for patterns or trends of problems and their probable sources.

5. Select several key documents and check the correctness of the inputs and outputs as compared to existing procedures.

From: J. A. Burgess, Auditing an Engineering Organization, *Mechanical Engineering*, Sept. 1979, American Society of Mechanical Engineers, New York.

TABLE 15.3 Audit Report Contents

Purpose and scope of audit.

The date(s) of the audit.

A summary listing of areas/activities audited (including those that were found acceptable).

Names of the audit team members and their organizational affiliation.

A listing of the governing requirements which were used during the audit (policies, procedures, regulations, etc.).

A summary description of the results and findings from the audit. Identify all areas or topics which were examined. List the related findings and observations that pertain to a specific area or task. If no problems were found, make a statement to that effect. (Depending on the complexity or importance of the findings, it may also be appropriate to provide recommendations or suggestions on how the deficiencies noted should be corrected.)

A request to the organization audited to provide a plan for correcting the nonconformances listed in the report.

From: J. A. Burgess, Auditing an Engineering Organization, *Mechanical Engineering*, Sept. 1979, American Society of Mechanical Engineers, New York.

should be discussed, and the group that was audited should have the opportunity to clear up any misunderstandings or ask questions about the audit results.

Nevertheless, simply telling the representatives of the group audited is not enough. The results of the audit should be formally documented in a written report and provided to engineering management to inform them of the auditors' conclusions.

15.2.3 Reporting the Results

The leader of the audit team is responsible to see that the audit report is written and issued. The report must describe the audit results in sufficient detail to permit responsible management to understand the problems and take the necessary corrective action. It should be issued within approximately 30 days of completion of the audit. Table 15.3 describes the type of information that should be included in the audit report.

Remember that the report is for management. Managers want to know if there are any problems. If so, what's wrong? Summarize what

QUALITY ASSURANCE AUDIT REPORT
ABC Company
Middletown, U.S.A.

Date of audit: June 22-23, 1984

Scope of audit: This QA audit examined the adequacy of the design
                        assurance program.

Applicable
  requirements: Engineering department procedures manual.

Audit team: Team Leader: R. Joseph, Manager QA
                        Team Member: C. Nichols, QC/Reliability
                                      Engineer

Audit details and conclusions

1. *Specification preparation and approval*

   Five materials and process specifications were reviewed. The
   specifications were well written, appeared to cover the topics
   thoroughly, tolerances were specified and appear to be reasonable, and the specifications were signed off in accordance with
   the department procedures requirements.
   No discrepancies were noted.

2. *Outline drawings*

   All outline drawings are to be signed by the design engineer and
   the section supervisor, according to Procedure No. 3.6.5. The
   audit team reviewed outline drawings for six projects and found
   four of the drawings were signed only by the design engineer.
   Finding: Lack of approval of outline drawings by the section
   supervisor is a nonconformance with program requirements.

3. *Hand calculations in the designers' files*

   Although there are no procedural requirements defining what
   should be included, or how it should be prepared, the auditors
   noted conditions, which, in their personal opinion, did not follow
   good engineering practices.
   Observation: Hand calculations generally are neither signed nor
   initialed by the originator and often are not dated. In many instances the calculations were not titled and not identified to the
   applicable component or project.

FIGURE 15.1 Sample of a narrative audit report.

| ORGANIZATION AUDITED
                  Mechanical Design | SHEET ___1___ OF ___5___ |

STATEMENT OF REQUIREMENT AND SOURCE

Design calculations to be reviewed and signed off by qualified verifier
(Eng. Proced. EP16)

AUDIT FINDINGS/OBSERVATIONS

1. Heat transfer calculations for FDL 1039 Compressor dated Jan. 26,
   1983 show no evidence of design verification

2. Final design report for RXL Turbine signed by author only.

3. Stress calculations for Aft Turbine Wheel for BCA Project show no
   evidence of verification

| DATE OF AUDIT:
   June 9, 1983 | PREPARED BY:
   Jack Warren | APPROVED BY:
   Amy Kemp |

FIGURE 15.2 Sample audit report sheet.

you found and identify any requirements which are not being met. Be
concise but clear. Separate fact from opinion. Figure 15.1 illustrates
a portion of a sample audit report, showing a narrative style that can
be used. Figure 15.2 shows an example of an Audit Finding Sheet
that represents another way of reporting audit results. It is the author's opinion that the narrative style is the most effective.

Copies of the final report should be distributed to those persons who are responsible for the management of the engineering organization and have a need to know, including the chief engineer, engineering department manager, or Vice-President of Engineering. However, the audit report should not be given indiscriminately to other departments. A good working guideline is to limit the distribution to the management of the group audited and to the management of the group responsible for performing the audit.

15.2.4 Followup and Closeout

After the audit report is issued, the management of the groups audited are expected to investigate the findings and conclusions and develop appropriate plans for corrective action. A written response to the audit report should then be prepared and submitted to Engineering Management, including the person that requested the audit be performed. The response should define what has been, or will be, done to correct the problems and when the corrective action will be accomplished. This should typically be issued within about 30 days after the audit report is received. This time limit is to assure that corrective action can be initiated while the information is fresh in everyone's mind. For resolving complex or long-term problems, a plan of action and implementation schedule may be all that can be specified in the response document.

When the response document is issued, the audit team leader should review it promptly and verify that corrective action is planned or has been taken on all deficient conditions reported.

It is also a good idea for the audit team leader to issue an evaluation report, documenting the acceptability of each proposed corrective action and the timeliness of planned actions. Such a report should be issued within two weeks of receiving the audit response and should be distributed to the same persons who received the original audit report. This closes the loop on the audit process by acknowledging completion of corrective actions and identifies any areas where further management attention is needed. It also can flag items requiring future follow-up.

15.3 AREAS FOR INVESTIGATION

There are a number of areas in an engineering department which should be audited periodically. Although the items listed below may not apply in all cases, many of these will be common to most engineering departments. The subsequent paragraphs describe the areas and suggest points of interest to examine during an audit. The reader is again referred to Appendix 3 for additional items to consider.

15.3.1 Basis for Design

The starting point for the design process is typically a customer contract or a marketing specification. This may describe either standard
or special requirements which must be factored into the design of the
product. Since customer dissatisfaction will normally result if the
equipment is not delivered in accordance with the customer's order requirements, this is an area that should be checked frequently. Proper
translation of customer and marketing requirements into engineering
drawings, specifications, instructions and related documents should be
investigated. Look for specifics in the ordering requirements and then
attempt to verify that these same requirements have been properly incorporated by Engineering.

Another facet of the customer contract requirements can be
checked by examining samples of the final product that is about to be
shipped. A few spot checks may quickly show whether or not the selected customer requirements such as finish, trim, connection points,
etc. are being included in the engineering documents and correctly incorporated into the product.

15.3.2 Drawings and Specifications

The output of most engineering groups consists of engineering drawings and specifications. Many engineering departments have established standards and procedures which govern the preparation, use
and control of these documents. Such requirements as content, part
identification, document numbering, review/signoff authorization for
use in specific applications and the control of changes to engineering
drawings and specifications should be reviewed. Also, see if the correct revisions of the documents are being used.

Pay particular attention to the correctness and completeness of the
top assembly drawings, outline drawings and wiring diagrams. These
drawings are especially important for the proper assembly, installation
and use of the product.

15.3.3 Design Analysis

Many engineering organizations must perform various engineering analysis of their designs. This includes structural, thermal, fluid flow,
etc. The audit should investigate whether proper methods are being
used for a specific application and if the results are adequate to support the design. The analysis should also be checked to assure that
proper sizes, loads, etc. are used which are consistent with the final
design requirements. Verify the calculations really apply to the design that was authorized for manufacture and delivery to the customer.
It is not uncommon to find that analysis has been performed on an

earlier design than is now being built and is no longer applicable or that special customer requirements were not incorporated into the analysis.

Calculations should be checked to see they define what project or component the analysis applies to, that assumptions are appropriate, proper methods are used, and the units of measure are stated. The calculations should also be signed and dated.

Another area worthy of investigation is the preparation, checking and application of computer programs used in design analysis. Determine if the program used is appropriate for the application and the results appear to be reasonable. Do not assume that just because the results came from a computer that the answers are correct.

15.3.4 Test Programs

Many products must be tested prior to delivery to the customer. In some cases, these tests are of an engineering nature and are performed on prototype designs. Audits should examine test programs to determine if the design that was tested is consistent with the design that is being delivered to the customer. It is not uncommon to find that the adequacy of last-minute design changes or added design features have not been verified. Again, the customer requirements should be examined to determine if they have any impact on the test program and if these were properly incorporated in the tests.

The results from the test program should be documented in some form of data sheets or test reports to preserve the data for future use or reference. These records should be examined to see if the results are consistent with the specified design parameters and assumptions used in the design. Do the results prove the adequacy of the design? Also look for evidence that the test data is traceable to the specific component or project.

Another problem occurs in test records when anomalies or unexpected failures occurred during testing. Look for evidence that these problems have been eliminated through re-design or actions were taken to correct misunderstandings of the original requirements. The records should show that the problem areas were resolved.

15.3.5 Engineering Changes

Engineering changes are a frequent source of difficulty in most engineering organizations. It is common to find that an engineering change has only been partially incorporated into the design effort. It may be that the customer requirements were not disseminated to all parties involved, or that the change was not recognized as affecting work that had been previously performed on that particular product. One good technique is to select a particular engineering change and then follow it through the complete design process. Verify that all necessary

actions have been taken and that all appropriate organizations affected
by the change have been properly advised. Look at such areas as
supporting analysis, test efforts, instruction manuals, and spare parts
data.

15.3.6 Control of Nonconformances

On occasion a nonconformance from the design requirements will occur.
This may be in the engineering phase or in the manufacturing cycle.
All such nonconformances should be controlled in a predetermined manner, and Engineering should have a major role in the decision-making
process. Individual nonconformances from the original design may or
may not be significant, but it is important to have a controlled process
for identifying, documenting, and providing final disposition of the
nonconformance. The system should provide assurance that the final
product is adequate for its intended use. Some of these nonconformances may be handled locally. Others require higher management or
customer approval prior to proceeding further with the design. It is
especially important to have evidence of the decision to proceed with
the nonconformance in light of the product liability concerns that are
now widely apparent in the industry. An engineering audit should examine how well nonconformances are controlled.

15.4 ETHICS FOR AUDITOR CONDUCT

To be effective, auditors must perform their work in an ethical manner. Otherwise their credibility and objectivity may be seriously
questioned. Ethics in the auditing sense is defined as "the rules of
conduct which are recognized and accepted as fair and reasonable by
a group of persons who are subjected to the audit." Several guidelines
are provided in the following sections to assist auditors in recognizing
the kinds of actions and conduct which are generally considered as
ethical.

15.4.1 Role of the Auditor

The person selected as the auditor should normally possess a broad
experience and background in the particular field of engineering that
he will be auditing. An auditor needs a knowledge of modern technology and managerial controls to permit them to make appropriate judg-
ments of existing conditions and of the effectiveness of the engineering operation. They should be familiar with, and sensitive to, those
factors which contribute to the technical excellence required by the
company to be applied to its product. Auditors have the obligation to
examine, understand, and report to management the conditions as they

find them, including any apparent deficiencies or weaknesses in the system, the product, or the method of managerial control.

And, finally, auditors need to be thick-skinned, because they may be looked upon with suspicion, apprehension, or possibly even with disdain. They seldom win any popularity contests. However, if they seek facts objectively, they can provide a valuable service to their company.

15.4.2 Objectivity

The auditor must strive to be objective at all times when conducting and reporting the audit. During the course of an audit, the auditor will encounter ideas, opinions, suggestions and facts, although it will not always be obvious which is which. They should not be unduly influenced by comments from persons they interview. The auditors must be fair in their investigations by seeking factual data and observing actual conditions. In addition, they should avoid pre-judging a situation due to personal preferences or biases.

In the event an apparent deficiency is identified, the auditor must continue the investigation in sufficient depth to determine if it is a chance occurrence, a simple error, or an actual shortcoming in the management control process. They should seek to find out *what* went wrong, not *who* went wrong. An auditor would be well advised to continue the investigation of an area if the facts seem fuzzy. They should avoid jumping to conclusions when factual evidence of a problem is sparse. Very often there is more than meets the eye in understanding why things were done in a particular way. It is possible that an unusual approach may be just as valid as a more familiar method. Get the facts.

15.4.3 Compliance with Local Customs

The auditors should comply closely with the customs of the organization which they are investigating. This includes compliance with normal working hours, observance of established lunch periods, and being considerate of peoples' other time commitments when scheduling and conducting review meetings and interviews (when in Rome, do as the Romans do). Actions which are seen as either superficial or overbearing will immediately undermine their ability to do the job.

While interviewing or working with persons during the audit, each auditor should always follow the appropriate courtesies and behavior which they would expect to find in a business operation. The auditor should treat those being audited with respect and conduct themselves in a manner which would be acceptable to them if they had to reverse places.

15.4.4 Impersonal Reporting

Both verbal and written audit results should be reported to management in an impersonal manner. The auditor should avoid finger-pointing or placing blame on individuals for any conditions which he believes are less than adequate. The competence or personal traits of individuals should not be questioned or reflected in an audit report. One way to assure such impartiality is for the auditor to have the management of the group audited read the draft report prior to issue. Ask them to review it for objectivity and correctness of fact. The auditor should be prepared to take full responsibility for the conclusions and be able to substantiate the findings with facts or specific occurrences if challenged.

15.5 IMPROVING THE SYSTEM

An audit is not an end in itself. It is simply a means of identifying where there appears to be problems in the management control process. Thus, it is engineering management's responsibility to review the results from the audit and determine what, if any, actions need to be taken.

15.5.1 Taking Action

It is not uncommon for the management of the audited group to be very defensive about any problems reported by the auditors. They may contend that the auditor is overstating the significance of the problem or just doesn't understand the real situation. However, to get the most from the process, management must be prepared and willing to make an objective review of the matter and to take corrective action to resolve shortcomings identified during the audit. Otherwise, the time and manpower spent on the audit was wasted, and management's credibility will be reduced if identified problems are ignored.

It is common for experienced auditors to list problems in one of two categories—Findings and Observations. A finding normally is a deviation from a specific requirement. On the other hand, an observation is a condition which, in the auditors' opinion, will eventually cause problems if not corrected, although no actual deviations were discovered during the audit. Engineering management should assign responsibility within the Engineering Department to investigate both findings and observations. The specific items which were in nonconformance should be corrected, and the conditions or root causes which allowed the nonconformance to occur in the first place should be identified and actions taken to prevent it from happening again. The actions needed to correct management systems problems are similar to the type of actions described in Chapter 11 to correct hardware nonconformances.

The observations represent a more subtle set of problems. Engineering management needs to examine these critically to decide whether the concerns are truly valid or are simply due to the auditor's preferences or possibly limited knowledge or experience in this particular area. However, don't just pass over such items, especially if the auditor's opinions and credibility are generally respected in the organization. Consider all such observations as an early warning of approaching problems and act accordingly to avoid the difficulty later.

When considering what action to take to resolve a finding or an observation, recognize that telling the employees not to make that error again is seldom sufficient. Almost always, it takes a change in the methods or in the checks and balances to achieve lasting performance improvement. The necessary action may consist of additional training, a change in the existing procedures, reassignment of responsibilities, the introduction of a new method, etc. Further, some means of monitoring the revised process should be implemented to see if the corrective actions actually resolve the problem. This needs to include a feedback reporting system to see if, and how many of, the same problems recur in a reasonable period of time after corrective action was taken. If no new occurrences are discovered during a reasonable period of time, the monitoring and reporting can be discontinued.

15.5.2 Audit Timing

You may ask: "When is an audit like this needed?" and "How often should it be repeated?" If a systems audit has never been done or hasn't been done for several years, it is appropriate to perform such an audit to establish a benchmark of current performance. In the event the audit reveals several problems, a followup audit should be conducted within three to six months after corrective action is implemented. This will provide timely feedback on whether or not the shortcomings in the system have been corrected.

There are several other times when an audit may be needed. These are:

After major changes are made in the organization or in the methods and procedures for doing the work.

When product feedback indicates a deterioration in the performance, reliability or safety of the product.

After a disaster or crisis situation, e.g., fire, flood, merger, financial hard times, which may alter the methods of operation.

In regard to audit frequency, some government contracts require annual audits. However, if it is not required by contract or industry practice, a good rule of thumb is to conduct audits every two - three years, if management is serious about maintaining an effective design assurance program. A three-year cycle is usually short enough to prevent small but typical problems from developing into serious situations and long enough to be resource-effective.

The mere act of conducting periodic audits goes a long way towards stimulating compliance with established methods. The department personnel know that random checks will be made and tend to act accordingly. Some persons may not like the thought of their work being audited. But in the long run, a program of objective audits will be regarded as a systematic approach for maintaining a high level of performance and integrity in the Engineering Department.

15.6 SUMMARY

Auditing techniques can be used to examine the effectiveness of an engineering organization and its design assurance efforts. Periodic audits examine how the work is done and to what extent the work methods satisfy the applicable requirements. Properly planned and executed audits provide management with useful insights into existing and potential problems. Management is then in position to take necessary corrective action to resolve existing problems and prevent recurrences. By using the audit approach, management conveys a message of being serious about having the work performed correctly the first time. It is an important tool in any defect prevention program.

16

Design Assurance in the Future

16.1   Engineering Management Responsibilities   276

16.2   Drawings   276

16.3   Specifications   277

16.4   Configuration Control   277

16.5   Design Methods and Analysis   277

16.6   Engineering Software   278

16.7   Product Testing   278

16.8   Design Reviews   278

16.9   Statistical Tools   278

16.10   Control of Nonconformances   279

16.11   Engineering Records and Supporting Documentation   279

16.12   Reliability Improvement   279

16.13   Conclusions   280

The purpose of design assurance is to improve the chances of success and reduce the risk of failure when introducing new designs. This effort will apply equally well in years to come as it does now. Many factors will be present which will make it important for the new designs to be right the first time. These include: the rapid advances in technology, the high cost of a lengthy product development effort, the time crunch of converting a new idea into a marketable product, the desire to make new products which have greater capabilities but are simpler to use, the spread of competition in the world, and the hazards and consequences of product defects.

These factors tend to work at cross-purposes and become formid-
able challenges to engineering organizations. To be successful, those
responsible for product design need both good ideas and good methods
for control. The ideas must come through the creative processes in the
future as they have in the past, and the control system should not in-
hibit creativity. But control will be needed. Such methods must be
developed and applied in a manner consistent with the demands of the
future. Let's look at how various design assurance methods and practices may be affected in years to come.

16.1 ENGINEERING MANAGEMENT RESPONSIBILITIES

In the future, as in the present, engineering management must set
high standards in the on-going pursuit of excellence. Management has
the obligation to establish policy which emphasizes the importance of
quality in design. This includes the implementation of effective design
assurance methods and practices. Further, engineering management
will be expected to verify by review and by audit that the design assurance program is in place, is followed and works effectively. Anything less will be seen by their organization as lack of commitment to
quality. Thus, the onus for leadership in the pursuit of excellence
will remain on engineering management.

16.2 DRAWINGS

Over the next several years, there will be much greater use of CAD/
CAM in defining the product. This will largely eliminate the use of
hardcopy drawings. The engineering output will go directly by digital
electronics to automated manufacturing equipment. Instead of drawings, the design will be viewed primarily on computer displays.

Although this will change the media for examining the designs, it
will not lessen the need for adequate review of the designs. In fact,
it places even greater importance on the front-end aspects of engineering, i.e., the definition of applicable requirements and the integration
of design concepts to satisfy those requirements.

In years to come, few drawings will actually be signed off. However, the designer will still be responsible for providing complete and
correct information. Therefore, other design assurance methods will
be needed to assure the proper level of quality is achieved in the drawings. It is expected that greater emphasis will be placed on having
knowledgeable persons outside of engineering participate regularly in
reviewing and concurring with the conceptual and preliminary designs.
Afterwards, the detailed designs will be completed and released through
the electronic media for production use.

16.3 SPECIFICATIONS

Engineering specifications will continue to be applied much as is done presently. The primary change will be in the methods used for publication and data storage. Computer-oriented devices will further advance the science of word processing and recordkeeping. In the plant, specifications will be viewed on computer terminal screens. Although hard copy specifications will still be used for most outside procurement, key suppliers will receive and exchange technical data with the buyer directly through computer networks.

Engineering will be responsible for the accuracy and completeness of the data. However, to get the best possible specifications, they will seek specialty inputs from other contributing groups.

16.4 CONFIGURATION CONTROL

The processing and control of changes will be integrated into the electronic drawing system. The system will routinely compile the changes, modify the design data, update the supporting technical documentation, and maintain the design baseline integrity. However, it will not replace the human judgement in making the approval decisions. People will still have to decide: "Is this the proper thing to do?" Although the automated systems may provide relief from many of the manual tasks and provide a wealth of information about the impact of the change, (e.g., where it will apply, what must be changed, what all is involved, what it will cost, etc.), the final decision to accept or reject the proposed change will remain a management responsibility.

16.5 DESIGN METHODS AND ANALYSIS

This is another area which will be computerized and automated extensively. Analytical methods will grow in importance as modeling techniques are improved. It will be used to replace much of the design-and-test methodology of the past. Comprehensive analysis will be a cost-effective alternative to many test programs.

Interactive (man-machine) design and analysis will be commonplace and will greatly improve engineering productivity. But it will also present a sizeable challenge for assuring the quality of design. The validity of the output will continue to be dependent upon the validity of the logic and input. The limited visibility and restricted access of computer-aided design methods makes it very difficult to check the work in the middle or at the end. Consequently, there will be a pressing need for learning and applying the best possible methods for software design and control.

16.6 ENGINEERING SOFTWARE

New computers and operating systems will stimulate the growth of interactive man-machine systems. Nearly all engineers will do some programming, and the number of engineering software programs in use
will expand exponentially. However, the old adage of "Garbage In -
Garbage Out" will still apply. Engineering management must insist
that their engineers learn and apply the best methods of structured
programming. Although new computer languages and methodology will
provide numerous checks and balances, it is still imperative that the
logic of the program be complete and correct for the task to be addressed. The world of engineered products will be heavily dependent upon
the quality of software design and its proper application.

16.7 PRODUCT TESTING

Testing is another area which will be revolutionized by automation in
years to come. Automated data acquisition and analysis will be applied
extensively, and automated test equipment will be integrated with the
CAD/CAM systems. Testing requirements will be developed in the design process, transmitted automatically to the test equipment and
stored electronically until needed. Later, the tests will be performed
automatically; data gathered, analyzed and reported; and the products
will be accepted or rejected. Those rejected will be tagged to show
what was wrong and what must be done to correct the anomaly. But
with the advent of automated testing, it will be a challenge to assure
that quality of test instructions and results is consistent with the need.

16.8 DESIGN REVIEWS

Design reviews will be important design assurance tools for the future.
They will be particularly significant in the establishment and validation
of the design requirements and in the selection of the design concepts
to be used. Good engineering judgement based on real-world experience and supported by powerful analytical tools will be the foundation
for decision-making. But two heads will still be better than one.

Electronic design methods will greatly improve productivity and
compress the engineering cycle. However, it will not automatically improve the quality of design decisions. As a result, design reviews by
multi-disciplined teams will be needed to provide valuable inputs for
these decisions.

16.9 STATISTICAL TOOLS

There will be a widespread application of statistical techniques in the
design and development cycle in the future. Statistical methods will be

integrated into design and analysis software and will be used routinely to assess the impact of variability and for economic design. Nearly all engineers, at all levels, will need to be trained in the application of statistics and in the interpretation of statistically-significant results.

16.10 CONTROL OF NONCONFORMANCES

Decisions regarding the resolution of nonconformances will continue to be a part of the designer's job. The major change will be the extensive use of computers to explore the impact of the nonconformance and assist the engineer in making the decision. This system will also provide a data base for trend analysis of recurring problems. However, the adequacy of such evaluation depends on the validity and completeness of the program logic and the correctness of the inputs. As usual, the data must be right, or the decisions won't be.

16.11 ENGINEERING RECORDS AND SUPPORTING DOCUMENTATION

The on-going advances in engineering methods and technology will not reduce the need for records, but their form will be changed drastically. More and more engineering data will be stored electronically and less of it will be the traditional paper copies. CAD/CAM systems, integrated with word processing equipment, will produce, store and retrieve a large portion of the engineering documentation. Thus, the planning for records will have to be done as a part of the design process. To a large extent, this will reduce the drudgery and improve the quality of records in many engineering departments. Records will tend to be retained in a central location. This should change the patterns of the past where each engineer kept a few scribbled notes about the job somewhere in his desk.

The use of electronic records appears as a bright spot in the future, since most engineers dislike the recordkeeping function with a passion.

16.12 RELIABILITY IMPROVEMENT

The need for reliability improvement will be present as long as someone builds a product in a competitive marketplace. The technology advances will provide tools for better design, analyses, and data collection. Further, it will provide new capabilities for exploring the many "what if . . ." questions of engineering more quickly and economically than in the past. However, these new methods still will not guarantee perfection in an imperfect world. Consequently, there will be an ongoing need to monitor a product's performance thoroughly. This must include data from the factory and the field to know how well the product satisfies the customer's needs and requirements. Reliability engineering techniques will play an important role in the future.

16.13 CONCLUSIONS

And, now, one final word about design assurance in the future. The methods and practices may change as more automation is applied in the engineering process. But the purpose will remain the same as today—to reduce the risks associated with introducing new designs and to assure these designs meet their applicable requirements.

The pursuit of excellence in engineering is a never-ending journey. Design assurance techniques provide some solid stepping stones along the way.

Appendixes

APPENDIX 1 Equipment Specification Contents (Reference U.S. Air
Force Systems Comand, Data Item C-2)

1. Scope:	Intended general use of the item.
2. Applicable documents:	List those documents (specifications, standards, bulletins, manuals, etc.) which are applicable to paragraphs within this specification.
3. Requirements:	This section contains performance and design requirements which are suitable for proof during test. Include design constraints.
3.1 Performance:	
3.1.1 Functional characteristics:	List limiting functional characteristics to include those which are established by analysis as well as those which are determined by design, as listed below.
3.1.1.1 Primary performance characteristics:	Include the products of analysis in terms which do not preselect design solutions. Include quantitative terms.
3.1.1.2 Secondary performance characteristics:	Recorded after a basic design approach has been established. Include those parameters which are not necessarily mission critical, but which must be specified to properly constrain a complete design, such as maximum continuous operating time at emergency rated power. Stated in quantitative terms.
3.2.1 Operability:	Include performance requirements which are general measures of efficiency, as listed below.
3.2.2.1 Reliability:	Stated in quantitative terms, such as measures of availability, mean time to failure, duration of downtime, etc.
3.1.2.2 Maintainability:	Include scheduled organization maintenance per storage or

		operating hour. Describe ease of service. Define requirements for access doors, built-in tools, self-test capability, inspection windows, test jacks, sealed and life warranty components, etc.
3.1.2.3	Useful life:	Include requirements for shelf-life, storage, as well as operating life cycle and combinations thereof.
3.1.2.4	Environmental:	State the environment which the item is to withstand, e.g., temperature, pressure, humidity, etc.
3.1.2.5	Transportability:	Restrictions on maximum dimensions and weight.
3.1.2.6	Human performance:	Specify requirements related to user operation, such as human size, strength, color recognition, fool-proof assembly, etc.
3.1.2.7	Safety:	Requirements which must be included to preclude or limit hazard to personnel and/or equipment. Refer to hazards in assembly, disassembly, test, transport, storage, operations, and/or maintenance. Fail safe and emergency restrictions shall be specified herein. This shall include applicable requirements for interlocks, emergency and standby circuits, etc.

3.2 Equipment definition:

3.2.1	Interface requirements:	Specify, either directly or by reference, relationship to other equipment. Include schematic arrangement and/or suitable data.
3.2.1.1	Schematic arrangement:	Show relationship to other equipment/facilities. This paragraph shall incorporate a schematic arrangement (diagram), inboard profile, or equivalent engineering drawing.

3.2.1.2 Detailed interface
       definition:
                           Express in quantitative terms with tolerances. Include mechanical and electrical relationships. Refer to common datum plane or point.

3.2.2 Component identification: List in categories because of engineering or supply significance. Include both mechanical assemblies as well as materials such as hydraulic fluid, fuels, bonding agents, etc.

3.3 Design and construction:

3.3.1 General design features: Specify physical characteristics such as size, weight, shape, individual critical dimensions, etc. Requirements may be descriptive or expressed in quantitative terms. All requirements should be verifiable by inspection.

3.3.2 Selection of specifications and standards: List requirements, criteria, constraints, pertinent to selection and imposition of federal-military-contractor specifications and standards. (Refer to MIL-STD-143.)

3.3.3 Materials parts and processes   Specify requirements for, or prohibit the use of, individual and/or types of materials, parts, and processes. Refer to pertinent specifications.

3.3.4 Standard and commercial parts: Describe criteria for selection.

3.3.5 Moisture and fungus resistance: State requirements in terms of humidity, time, temperature, etc.

3.3.6 Corrosion of metal parts: State requirements for use of protective coatings, the method for controlling electrolytic corrosion, etc.

3.3.7 Interchangeability and replaceability: Define lowest level of assembly at which repair parts will be stocked. (Refer to DOD-STD-100.)

3.3.8 Workmanship:	State general requirements related to fabrication.
3.3.9 Electromagnetic interference:	State in terms of the environment which the item must accept and the environment which it generates.
3.3.10 Identification and marking:	Include coding for wiring, plumbing, identification of hazardous conditions, explosive components, nameplate requirements, etc.
3.3.11 Storage:	Include maximum storage durations, storage environment, restrictions pertaining to maintenance while in storage, etc.
4. Quality assurance provisions:	Requirements for formal verification of the design performance and construction of the item. Specify a verification requirement and method for each requirement in Section 3.
5. Preparation for delivery:	Specify what must be done prior to delivery, such as add electrolyte, pressurize, etc.
6. Notes:	This section shall include information which is stated for administrative purposes. Not a part of the specification in the contractual sense. Include background information or rationale which will be of assistance in understanding the specification.
7.	Used for special situations only.
8.	Used for special situations only.
9.	Used for special situations only.
10.	Specify requirements which are contractually a part of the specification but are limited to requirements of a temporary nature—that is, interim performance for early test models.

APPENDIX 2 Sample Design Review Checklist

Each of the following items should be considered. Negative respon-
sives should be identified and evaluated for possible corrective
action.

1. Does the design meet the performance requirements for the
   application?

   a. Normal operating, steady state
   b. Normal operating transients
   c. Start up/shutdown
   d. Emergency or overload
   e. Strength
   f. Motion/travel
   g. Size/weight
   h. Duty cycle
   i. Useful life
   j. Operating times

2. Is the design capable of meeting the environmental requirements?

   a. Loads (mechanical, electrical, thermal, etc.)
   b. Temperatures (transportation, operating, operative storage,
      inoperative storage)
   c. Humidity
   d. Shock/vibration (operation, transportation)
   e. Corrosive ambients (salt air, sea water, contact chemicals,
      etc.)
   f. Immersion (water, oil, cleaning agents, etc.)
   g. Pressure/vacuum
   h. Weather (sun, rain, snow, ice, wind)
   i. Acoustic ambient
   j. Magnetic field
   k. Radio interference
   l. Friction
   m. Galvanic corrosion

3. Is the design capable of meeting the cost objectives?

   a. Material
   b. Labor
   c. Tooling
   d. Value engineering analysis performed?

4. Does the design have a high probability of meeting the reliability requirements for the application?

   a. Stresses within acceptable limits
   b. Derating utilized

    c. Problems from similar designs corrected or avoided
    d. Failure modes from similar designs eliminated or avoided
    e. Proven parts used
    f. Satisfactory results from design integrity tests

5. Does the design meet the producibility requirements?

    a. Materials/processes defined clearly
    b. Optimum standardization of parts and materials
    c. Production problems from similar designs corrected or avoided
    d. Utilize vailable equipment/processes
    e. Inspection/test requirements defined
    f. Acceptance criteria defined with reasonable tolerances
    g. Assembly provisions defined
    h. Workmanship/finishes defined

6. Does the design satisfy applicable health/safety requirements?

    a. OSHA
    b. Insulation/interlock provisions
    c. Safe for intended use and foreseeable misuse
    d. Free of sharp edges, pinch points, etc.
    e. Adequate hazards protection/warnings

7. Does the design meet applicable maintenance/serviceability requirements?

    a. Access available for adjustment/repair
    b. Need for special tools/fixtures minimized
    c. Adequate protection provisions for safe servicing
    d. Installation/calibration/operation instructions available and clear

8. Does the design meet the aesthetic requirements?

    a. Appealing to the eye
    b. Color/finish
    c. Logical arrangement and size for human factors considerations
    d. Convenient and efficient arrangement and placement of controls and monitoring devices
    e. Proper blending/contrast to installed environment

APPENDIX 3 Guidelines for Auditing the Engineering Function

A. *Basic question*:

Does the Engineering Function provide the necessary control and verification to assure the adequacy of the product design?

B. *Audit technique*:

1. Obtain a set of procedures and/or instructions which apply to the Engineering group. Review these documents thoroughly to understand the scope of effort and the flow of information. Verify that procedures exist for controlling those basic engineering functions performed by the group, such as:

    a. Preparation and approval of drawings/specifications
    b. Verifying the adequacy of the design by analysis and/or test
    c. Controlling changes to the design
    d. Controlling nonconformances from the design requirements.

2. Discuss the functioning of the Engineering organization with the Engineering manager. Ask the manager to explain the basic procedures and the flow of information.

3. Select 4-5 design projects or shop orders to examine. Choose tasks which are representative of the scope of work. Select 1-2 projects that were completed within the past year and 1-2 projects that are presently active. (Be sure to select projects that have progressed sufficiently to see how things really work.)

4. Review the requirements received from Marketing (or customer). Note specific technical requirements, e.g., size, weight limits, performance levels, environmental conditions, test requirements, etc. Check to see that these specific requirements are incorporated into the product design process.

5. Select 4-5 drawings for two or three projects. Choose drawings that cover the range of the product (parts, subassemblies, final assembly, wiring diagrams, installation drawings, etc.). Include some drawings that the customer sees or uses. Determine if the drawings satisfy the stated requirements for preparation (format, data presentation, etc.) and approval (signoff).

6. Examine at least 8-10 drawing revisions. Where possible, investigate changes that affect other drawings or documents. Determine if the changes were made consistently. Verify that the proper approvals were obtained for each change.

7. Select 4-5 specifications covering materials and parts. Determine if the specifications satisfy the stated requirements for content and approval. Examine revisions for 4-5 specifications. Determine if the changes were made consistent with the requirements for the change. Verify that the proper approvals were obtained for each change.

8. Examine the process for releasing the engineering information for production/procurement use. Determine if the process gets the correct information to the proper persons in a timely manner. Look at the release of new design information and the release of revised design information.

9. Where applicable, examine 4-5 design calculation packages. Look for evidence that the design satisfies the specification requirements and is considered adequate for the intended use. Look for evidence of proper approval.

10. Where applicable, examine 4-5 examples where physical testing is used to establish design standards or verify design adequacy. Determine if the test results demonstrate that the design meets the specified requirements. Verify that any noted test anomalies are properly resolved.

11. Determine if the design that is tested is representative of the final design. Look for evidence of consistency in applying the results to a final design.

12. Look for examples of significant engineering changes. Follow 4-5 changes through to determine if the necessary design and verification packages were modified consistent with the change. Look at applicable drawings, specifications, calculations, test results, instruction books, spare parts data, etc.

13. Verify that the Engineering Function is actively involved in the process for deciding the appropriate action to be taken on product nonconformances from the engineering requirements. Determine if adequate management controls are in place for assuring that the nonconforming items are eliminated or are determined to be adequate "as is" or with a prescribed repair.

14. Look for evidence that the Engineering Function receives product performance feedback (discrepancies, failures, customer complaints, etc.) from the factory and the field and incorporates the results into the engineering process.

Conclusion:

Overall, are you satisfied that the Engineering Function adequately controls the design of products in a manner that reasonably assures that the specified requirements will be fulfilled? If not, why not?

Selected Readings

RELATED TEXTS

Cronstedt, Val (1961), *Engineering Management and Administration*, McGraw-Hill, New York.
> This book describes many of the practices for the administration and management of an engineering department. The chapter on engineering changes contains many good ideas and practices.

Dunn, Robert and Ullman, Richard (1982), *Quality Assurance for Computer Software*, McGraw-Hill, New York.
> The authors present a detailed explanation of controlling the quality of computer software. This is the first comprehensive work of this type.

Emrick, Norbert L. (1977), *Quality Control and Reliability*, 7th Ed., Industrial Press, New York.
> This is a good basic book on quality control. It contains several well-written chapters on reliability.

Feigenbaum, A. V. (1983), *Total Quality Control*, 3rd Ed., McGraw-Hill, New York.
> This is a classic book on the broad topic of Quality Control. It contains much valuable information, and the sections on New Design and statistical tools are easy to understand and use.

French, Thomas E. and Vierck, Charles J. (1953), *Engineering Drawing*, 8th Ed., McGraw-Hill, New York.
> This is a standard text for the training of drafting methods and techniques. It covers all fundamentals of drawings and their preparation.

Fuller, Don (1966), *Organizing, Planning and Scheduling for Engineering Operations*, Industrial Education Institute, Boston.
> This book provides many practical methods for use in operating an engineering department. It contains many helpful tips pertaining to the drafting function.

Grant, Eugene L. and Leavenworth, Richard S. (1980), *Statistical Quality Control*, 5th Ed., McGraw-Hill, New York.
This book is a well-organized authority document on all phases of statistical quality control. It covers the basic principles as well as application to practice.

Gresecke, Frederick E., Mitchell, Alva, Spencer, Henry C., Hill, Ivan L. and Loving, Robert O. (1969), *Engineering Graphics*, Macmillan, New York.
There is a good description and discussion of design and working drawings in this book.

Hicks, Tyler G. (1966), *Successful Engineering Management*, McGraw-Hill, New York.
This book provides many practical insights into the management of an engineering department. The section on control of engineering design touches on several design assurance methods.

Johnson, L. Marvin (1982), *Quality Assurance Program Evaluation*, Revised Ed., Stockton Trade Press, Santa Fe Springs, California.
This is probably the most comprehensive book available on the topic of quality assurance auditing.

Juran, J. M. and Gryna, Frank M. Jr. (1980), *Quality Planning and Analysis*, 2nd Ed., McGraw-Hill, New York.
This is an excellent text on all phases of quality assurance. It contains particularly good sections on statistical tools and reliability.

Kirkpatrick, Elwood G. (1970), *Quality Control for Managers and Engineers*, Wiley, New York.
This contains a good discussion of statistical tolerancing, in addition to many basic quality control and statistical topics.

Lloyd, David K. and Lipow, Myron (1977), *Reliability Management, Methods and Mathematics*, 2nd Ed., Published by authors, Redondo Beach, California.
This is excellent text on all facets of reliability. It contains theory, practice and practical examples, including a large chapter on software reliability.

MacNiece, E. H. (1953), *Industrial Specifications*, Wiley, New York.
This text covers the basics of preparing and using various types of specifications.

Myers, G. J. (1976), *Software Reliability*, Wiley, New York.
Many of the causes and effects of software problems are covered in this book. The author identifies various methods for predicting and measuring the reliability of large computer programs.

O'Connor, Patrick D. T. (1981), *Practical Reliability Engineering*, Heyden, London.
This is a very readable book and covers the topic very well. It explains most of the reliability tools in use today. There is also a good chapter on software reliability.

Philippakis, Andrew S. and Kazmier, Leonard J. (1982), *Advanced COBOL*, McGraw-Hill, New York.
> This provides a comprehensive discussion of the use of the COBOL language. It includes many useful programming techniques and tips for avoiding errors.

Samaras, Thomas T. (1975), *Engineering Graphics Desk Book*, Prentice-Hall, Englewood Cliffs, New Jersey.
> Many basic concepts and practices for the preparation and control of engineering drawings are included in this text.

Shooman, M. (1980), *Software Engineering*, McGraw-Hill, New York.
> The fundamentals of structured programming are explained thoroughly in this text. The author presents many programming techniques which contribute to avoiding errors in programming.

VerPlanck, D. W. and Teare, B. R. Jr. (1954), *Engineering Analysis*, Wiley, New York.
> There are many fundamental concepts and approaches to the analysis of engineering problems described in this book. It provides insights into the overall analysis process.

Walton, Thomas F. (1968), *Technical Manual Writing and Administration*, McGraw-Hill, New York.
> The process of planning and publishing technical manuals is described in this text. It also covers the managing of a technical writing group.

GOVERNMENT AND INDUSTRIAL SPECIFICATIONS AND STANDARDS

ANSI Std. N45.2.11 (1974), *Quality Assurance Requirements for the Design of Nuclear Power Plans*, American Society of Mechanical Engineers, New York.
> Defines a set of overall requirements for a design control system. It is one of the earliest such documents and presents a clear view of the requirements.

ANSI/IEEE Standard 730 (1981), *Standard for Software Quality Assurance Plans*, Institute of Electrical and Electronics Engineers, New York.
> Defines the requirements for preparing a management plan for controlling new computer software projects.

ATA Spec. No. 100 (1970), *Specification for Manufacturers' Technical Data*, Air Transport Association, Washington, D.C.
> Describes the requirements for preparing and publishing technical manuals for commercial aircraft.

DOD-STD-100 (1978), *Engineering Drawing Practices*, U.S. Department of Defense, Government Printing Office, Washington, D.C.
> The basic requirements for all types of drawings used for military contracts are identified and explained in this standard. These practices have wide acceptance throughout industry.

MIL-STD-480A (1978), *Configuration Control*, U.S. Department of Defense, Government Printing Office, Washington, D.C.

Establishes the major requirements for controlling changes to equipment produced under government contract.

MIL-STD-483 (1970), *Configuration Management Practices*, U.S. Department of Defense, Government of Printing Office, Washington, D.C.

This is a companion document to MIL-STD-480. It describes the overall operation of a change control system for equipment produced under government contract.

MIL-STD-499A (1974), *Systems Engineering Management*, U.S. Department of Defense, Government Printing Office, Washington, D.C.

Presents a disciplined approach for defining and controlling engineering activities for major government projects. It is broadly related to design control activities.

MIL-STD-785B (1980), *Reliability Program for Systems and Equipment*, U.S. Department of Defense, Government Printing Office, Washington, D.C.

Specifies the various work program elements for organizing and managing a reliability program. It contains many useful ideas that can be applied to non-defense products.

MIL-STD-781C (1977), *Reliability Design Qualification and Acceptance Tests*, U.S. Department of Defense, Government Printing Office, Washington, D.C.

The requirements for planning and performing reliability tests are defined in this standard. It specifies sample sizes, test frequencies, etc., to obtain valid results for demonstrating compliance with reliability requirements.

MIL-STD-1629A (1980), *Procedures for Performing Failure Mode, Effects and Criticality Analysis*, U.S. Department of Defense, Government Printing Office, Washington, D.C.

This is the basic document for conducting a FMECA. It describes the terms and methodology.

MIL-STD-1679 (1978), *Weapon System Software Development*, U.S. Department of Defense, Government Printing Office, Washington, D.C.

This standard defines the detailed requirements for providing computer programs under government contracts. It contains many useful explanations of what is needed and why. These practices are gaining broad acceptance in the computer world.

MIL-S-52779 (1974), *Software Quality Assurance Program Requirements*, U.S. Department of Defense, Government Printing Office, Washington, D.C.

Defines the elements of a program for managing the development of computer software under government contract. This standard can be used in conjunction with MIL-STD-1679.

MIL-S-83490 (1968), *Specifications, Types and Forms*, U.S. Department of Defense Government Printing Office, Washington, D.C.

> The various types and forms of specifications used for government contracts are described in this document.

Canadian Std. Z299.1 (1978), *Quality Assurance Program Requirements*, Canadian Standards Association, Rexdale, Ontario.

> This is a broad program requirements document. It includes a detailed section on Design Assurance program requirements.

Note: The U.S. government specifications and standards are available from:

Naval Publications and Forms Center, 5801 Tabor Avenue, Philadelphia, PA 19120

RELATED ARTICLES BY THE AUTHOR

Burgess, John A., File Now - Find Later, *Machine Design*, Penton/IPC, Cleveland, OH, April 28, 1966.

> This article describes the various aspects of filing, storage and retrieval of engineering records.

Burgess, John A., Making the Most of Design Reviews, *Machine Design*, Penton/IPC, Cleveland, Ohio, July 4, 1968, pp. 90–95.

> Various aspects of planning, performing and follow-up of design reviews are described. The article focuses on the use of formal design reviews to verify the adequacy of a design.

Burgess, John A., Organizing Design Problems, *Machine Design*, Penton/IPC, Cleveland, Ohio, Nov. 27, 1969, pp. 120–127.

> The process of conducting functional analysis is explained in this article. It shows how to define and document the requirements for product design.

Burgess, John A., Spotting Trouble Before It Happens, *Machine Design*, Penton/IPC, Cleveland, Ohio, Sept. 17, 1970, pp. 150–155.

> Conducting failure mode and fault tree analysis are explained in this article. These techniques apply to both reliability and safety evaluations.

Burgess, John A., Keep Track of Your Design Efforts, *Machine Design*, Penton/IPC, Cleveland, Ohio, July 8, 1971, pp. 66–69.

> Methods for documenting and filing engineering calculations are described in this article. It explains the use of a Design Release Memorandum and how to apply microfilm storage and computer indexing for this type of information.

Burgess, John A., Auditing an Engineering Operation, *Mechanical Engineering*, American Society of Mechanical Engineers, New York, September 1979.

> The basics of planning, performing and reporting an audit of an engineering department are presented. Also guidelines for specific areas to investigate and auditor conduct are included.

Burgess, John A., Quality Assurance for Computer Software, *Quality*, September 1979, pp. 110–112.

   The ingredients of a software quality assurance program is explained in this article. It provides an overview of the process.

Burgess, John A., Assuring the Quality of Design, *Machine Design*, Penton/IPC, Cleveland, Ohio, February 25, 1982, pp. 65–69.

   Many of the basic elements of design assurance are contained in this article. It was the trigger which led to the writing of this book.

Burgess, John A., Quality Assurance in Testing, *Quality*, Hitchcock, Wheaton, IL, January 1983, pp. 36–37.

   The elements for initiating and controlling a test program are described in this article. It covers test planning, performing the test and reporting the results.

Index

Acceptance tests:
  for hardware, 152-153
  for software, 138, 140
Alphabetical index, 208-210
American Society of Mech-
    anical Engineers, 2
Approval signature, 78
Assembly drawings, 38-41,
    132
  preferred practices for,
    39-40
Assessment report, 257
Audits, 259-274
  planning and preparing for,
    261-262
  conducting, 262-264
  reporting, 264-267, 272
  closing out, 267
  frequency of, 273-274
  areas for investigation, 267-
    270
Auditors:
  characteristics of, 261
  ethics for conduct, 270-272
Average, 188

Bathtub curve, 234
Bill of materials, 38-39, 86-87
Bimodal distribution, 187

Binominal distribution, 188-
    189
Block diagrams, 23, 46
  for reliability, 252-253
Book form drawings, 43-44, 46
Bottom-up design, 129-130

CAD/CAM, 61-62, 279
Calculation summary sheet, 116-
    118
Calibration, 159
  instructions for, 223
CAPL, 89-90
Cause and effect diagrams,
    240-241
CCB, 96-97, 246
Cell boundaries, 185
Central tendency, 188
Change control cycle, 51
Checklists:
  for calculations, 121
  for design, 28-29
  design review, 176
  design review chairman's,
    174
  for functional flow diagrams,
    27
  for reliability reviews, 247
Chi square method, 241

Class I-III changes, 92-93,
  96-99, 204
Code inspections, 145
Coding practices, 133-136
Compiler, 136
Component testing, 255
Computer file, 80-81
Computer programs:
  coding of, 132-136
  debugging, 136-138
  preferred constructs for,
    133-135
  structures of, 130-131
  verification of, 138-140
Configuration control, 84-103,
    277
  of engineering software, 142-
    145
Configuration Control Board,
    96-97, 246
Configuration identification,
    85-90
Configuration verification, 85,
    102-103
Consultants, 76
Controlled assembly parts
    list, 89-90
Corrective action, 204-205,
    244, 272-273
Critical failure mode list,
    249-251
Customer design reviews, 178-
    179
Customer specifications, 18

Debugging computer programs,
    136-138
  tools for, 137
Design:
  analysis, 114-122, 277
  baseline, 86
    for software, 143
  calculations, 115-122
    records of, 115-118

[Design]
  typical contents of, 116
  verification of, 118-121
  checklists, 28-29
  control, 2
  inspection, 119
  integration, 122
  manual, 114
  methods, 104-114
  requirements, 16-31
    external sources of, 18-
      20
    internal sources of, 20-21
Design assurance, 2-15
  definition of, 2
  policy, 5-6
  training, 7
  program, 10-15
Design review, 166-182
  chairman, 168, 173-178
    checklist for, 174
    duties of, 168, 173, 176-
      177, 178
  checklist, 176
  critique of, 178
  follow-up, 177-178
  in the future, 278
  key questions for, 176-177
  leadership techniques, 175
  open items, 177
  program, 170-171, 181-182
  for software, 145-146
Design specification, 48-50
Design verification:
  checklist, 121
  reviews, 179-180
Design verifier, 118-180
Design walk-through, 136
Desk-top review, 136, 145
Detail drawings, 36-38, 132
  characteristics of, 36-38
Dispersion of data, 188
Disputes, 72
Drawing:
  change requests, 57-60
  cycle, 47

[Drawing]
  numbering systems, 53-55
  release, 56-57
  requests, 48-49
Drawing control, 33-62, 276
Drawings, 33-62
  distribution of, 56-57
  preparation of, 46-50
  review and approval of, 50-
    53
  revision of, 57-62
  types of, 33-46
D-Spec, 48-50

Electrical diagrams, 45-46
Encyclopedia index, 210-211
Engineering changes:
  control of, 90-101
  classes of, 92-93, 96-99
  preparation of, 93-95
  review and approval of, 95-
    99
Engineering change, 90-92
  notice, 90-92
  order, 90
  proposal, 90
  request, 82, 90
Engineering department:
  areas to audit in, 267-270
Engineering management, 5-8
  leadership, 7
  responsibilities, 276, 277
Engineering procedures, 106-
    112
  index of, 112
  style of, 106, 108-110
  typical contents, 107
Engineering records, 207-220
  filing practices for, 211-217
  in the future, 279
  methods for indexing, 208-
    211
  physical security of, 220
  recommended indexing for,
    213

[Engineering records]
  retention of, 218-220
Engineering relationships:
  with Accounting, 10
  with Manufacturing, 9-10
  with Marketing, 8
  with Purchasing, 9
  with Quality Assurance, 10
Engineering software:
  acceptance testing of, 140-
    141
  characteristics of, 125-126
  coding practices, 133-136
  configuration control, 142-
    145
  debugging, 136-138
  design of, 128-132
  documentation, 141-142
  in the future, 278
  integration testing of, 139-
    140
  life cycle, 126
Ethics for auditors, 270-272
Excellence:
  pursuit of, 7-8, 276, 280

Failure analysis, 241-243,
    255
Failure criticality index, 248-
    250
Failure mode, effects and
    criticality analysis, 246-
    252
Filing practices, 211-217
Financial claims, 72
Flow gates, 26
FMECA, 246-252
Formal design reviews, 167-
    179
  conducting, 172-177
  general guidelines for, 167-
    169
  planning of, 169-172, 173
  for software, 146
  types of, 171

Frequency distributions, 184-
188
Functional analysis, 22-27, 129
Functional flow diagrams, 24-27
checklist for, 27
Functional requirements, 70

General arrangement drawings,
41-42
GO TO statements, 134-135

Hierarchical structure, 130-131
High order language, 135
Histograms, 185-186

Indexing, 208-211
alphabetical, 208-210
numerical, 210-211
rules for alphabetical, 209
types of alphabetical, 209
Industry standards, 18-19
Informal design reviews, 181
for software, 146
Installation instructions, 222
Instruction notes, 38, 40
Interviews during audits, 263
Introducing change, 14

J-distribution, 187
Juran, Dr. J. M., 3

Key word indexing, 116, 216

Layout drawings, 33-36, 132
characteristics of, 34

[Layout drawings]
controlling changes to, 34,
36
Library control, 143
Life tests, 256
Local customs, 271
Logic gates, 25

Maintenance instructions,
223
Marketing requirements, 21
Material card, 65, 67
Material Review Board, 203-
204
Material specifications, 64-66
content of, 65, 74
Materials of construction, 38
Mean time between failures,
241
Microfilm, 61, 219
Model tests, 151, 152
Module, 130-132
MRB, 203-204
MTBF, 241
Murphy's law, 85

Nameplate, 102
Network structure, 130-131
Nonconformances:
closeout and feedback of,
204-205
disposition of, 200, 203-
204
reporting of, 201-203
Nonconformance control, 198-
205
for software, 147-148
principles for, 198-200
Nonconformance report, 201-
203
Nonconformance tag, 201
Normal curve, 188-192
area under, 189, 191

Objectivity:
  in auditors, 271
  in verifiers, 118, 180
Operating instructions, 223
Operation and maintenance
      instructions, 222-228
  content of, 223-225
  guidelines for writing, 225-
      228
  illustrations for, 227
  verification of, 231

Parts lists, 87-90
  revision of, 88-90
  for replacement parts, 228-
      230
Problem reports, 237
  information flow for, 238
Process capability, 192-195
Process specifications, 65, 67-
      68
  content of, 69, 75
Product development require-
      ments, 21
Product feedback, 19-20
Product specifications, 69-71
Program design language, 132,
      137
Project files, 211, 217
Proof tests, 151-153
Proprietary information, 77
Proven parts, 245

Quality program requirements
      documents, 3

Range, 188, 190
Records:
  calculation, 115-118
  physical security of, 220
  retention times, 219

[Records]
  test, 154-163
Redundancy, 245
Reliability, 233-257
  assessment of, 256-257
  block diagrams, 252-253
  definition of, 234
  design tools for, 244, 254
  in the future, 279-280
  growth, 234
  mathematical models, 252, 254
  measuring and analyzing, 238-
      241, 256
  modeling for, 252-254
  reporting system for, 235-238
  reviews, 246
Repair disposition, 200
Repair and overhaul instruc-
      tions, 223
Replacement parts lists, 228-
      230
  parts identifiers for, 230
Reports:
  audit, 264-267
  reliability, 235-238, 256-257
  technical, 120-122
  test, 161-163
Requirements allocation sheets,
      29
Requirements review, 128
Revision column, 60-61, 99-
      100
Rework disposition, 200
Risk areas, 246
Role of auditors, 270-271

Scale model tests, 151
Sequential numbering, 210
Servicing instructions, 223
Sigma (σ), 188-189
Simplicity in design, 245
Skewed distribution, 186-
      187
Software:
  change log, 143-143

[Software]
  design:
    baseline, 143
    description, 130-131
    detailed, 132
    documentation package,
      141-142
    notebook, 128
    partitioning, 128-129
    requirements for, 128-129
    top-level, 129-131
  problem reporting, 146-148
  specification, 128, 141
  test plan, 131, 139-141
Source code listing, 141
Source control drawings, 42-
  43
Special processes, 38, 40
Specification:
  abstract, 80
  control, 63-83, 277
  cycle, 72-73
  format, 66, 68, 71, 72
  index, 79-80
  limits, 185, 193-194
  preparation, 72-76
  release, 79-81
  review and approval, 76-78
  revisions, 81-83
  signoff, 77-78
  writers, 72
  writing principles, 70, 72
Specification control drawings,
  42-43
Spread, 188
Standard design methods,
  113-114
Standard deviation, 189, 190
Standards writing, 1
Statistical:
  tolerancing, 195-196
  tools, 183-197, 278-279
Structured programming, 129
Subject files, 211, 215-216
  guidelines for, 215

Supporting documentation,
  221-232, 279
Systems testing, 255-256

Tally sheet, 184-185, 188
Technical reports, 120-122
  vertification of, 121-122
Technical societies, 19
Techniques of auditing, 263-
  264
Test:
  cycle, 154
  data:
    collection of, 159
    files, 163-164
    traceability of, 160
  declarations, 254-255
  planning, 156-157, 254
  predictions, 254
  procedures, 157-159, 160
    contents of, 158
    for software, 131
  program, 153
  reports:
    format for, 162
    guidelines for, 163
    preparation of, 161-163
    review and approval of,
      163
  requests, 155-156
Testing:
  in the future, 278
  of hardware, 151-165
  running the, 160-161
  of software, 138-140
Theory of equipment operation,
  224, 227
Tolerances:
  stackup of, 195-196
  standard drawing, 37
Top-down:
  design, 129-130
  testing, 139-140

Total quality control, 2
Trade associations, 19
Training test personnel, 159
Translation of requirements,
    27-30
Troubleshooting instructions,
    227-228

Use as is, 200, 204, 205

Variability, 160, 183, 185, 191
Verification tests:
    types of, 151

Wiring diagrams, 46

X-bar ($\overline{X}$), 188